Philo of Byzantium

PNEUMATICA

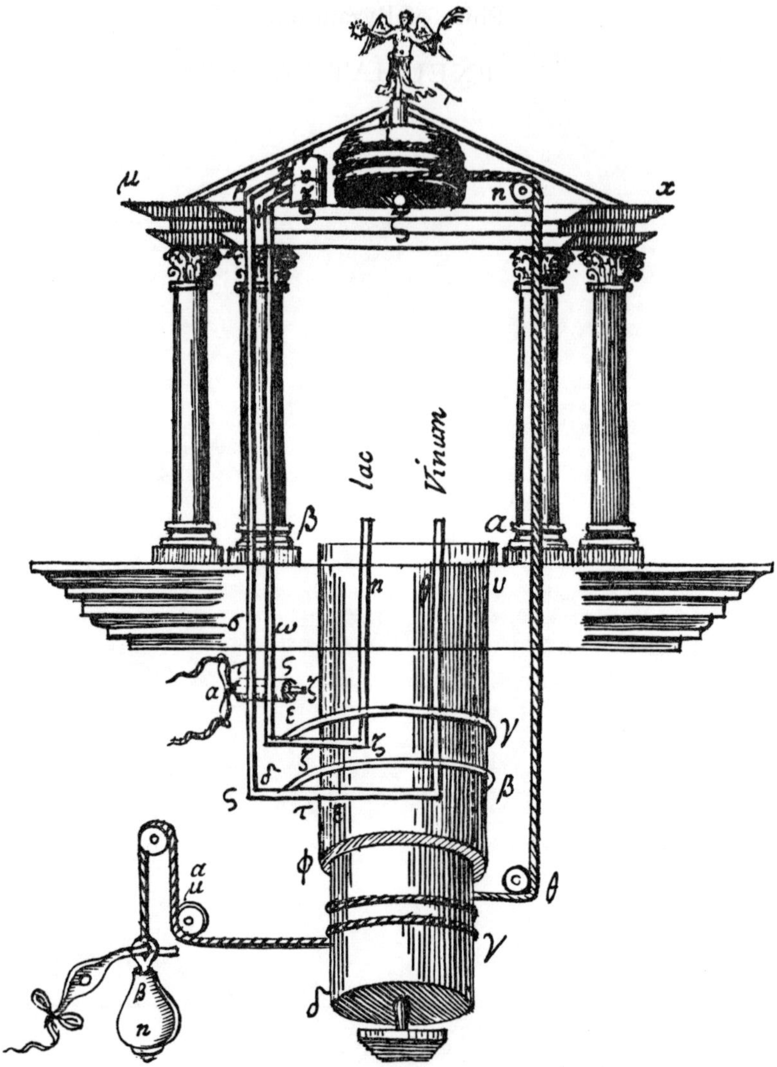

Figure 1. Ancient pneumatic automaton.
From a medieval copy, reproduced in 1693

Philo of Byzantium

PNEUMATICA

The First Treatise on Experimental Physics:
Western Version and Eastern Version

Facsimile and Transcript of the Latin Manuscript,
CLM 534, Bayerische Staatsbibliothek, Munich

Translation and Illustrations of the Arabic Manuscript,
A. S. 3713, Aya-Sofya, Istanbul

with Notes on other Manuscripts and Illustrations,
Historical Introduction, and Technical Commentary

by Frank David Prager

WIESBADEN 1974

DR. LUDWIG REICHERT VERLAG

Gesamtherstellung: Hubert & Co., Göttingen
Printed in Germany
ISBN 3-920153-32-4

Contents

List of Illustrations

Preface

The plan for this edition of Philo's *Pneumatica* was initiated in order to make the famous work available in English, to illustrate it with reproductions of authentic drawings, and to develop a fresh view of certain historic questions connected with it. As the work progressed I found it necessary to review additional questions of this kind. I then secured the assistance—not necessarily the concurrence—of two friends and associates. Both of them are professional historians, while I, starting with approaches and methods of patent law, take a more general interest in the history of discoveries and inventions.

One of my historian friends is Gustina Scaglia of New York. She contributed vitally to the study of documentary and stylistic problems. She also took the leading steps toward discovery and interpretation of the Italian Philo. As in former, joint efforts, Miss Scaglia's cooperation was a most rewarding experience.

My other friendly coworker is Alex Keller of Leicester. We had discussed some problems of ancient pneumatics before, and in due course we discussed this Philo edition. We did this during a number of years, by letters and occasional meetings. In many areas, the views of Dr. Keller have clarified my ideas.

Careful review of the Arabic Philo was needed. It was complicated by the fact that I do not read Arabic and could only use the French translation as a starting point. For the review of the Arabic text, Dr. Keller had the advice of his wife, who speaks and reads the language fluently. I had similar assistance from Mrs. N. Showker, teacher of languages in Johnson City, Tennessee, who was born in Lebanon.

Regarding certain problems of Greek and Latin texts, affecting the Philo record, I sought and received advice from others, particularly Fritz Krafft of Mainz and Martin Ostwald of Swarthmore. Their help is gratefully acknowledged.

I extend sincere thanks to Dr. Ludwig Reichert of Wiesbaden, for his great helpfulness and total unselfishness in making this edition possible.

Finally I wish to acknowledge the able assistance given by many librarians. In particular, pertinent bits of informations, as well as the required copies, were received from archivists in places where Philonian manuscripts are preserved: Cambridge, Florence, Geneva, Istanbul, Krakow, London, Munich, Oxford, Paris, and Vatican City. Thanks are due to the Bayerische Staatsbibliothek in Munich, the Ministry of Culture in Ankara, the Master and Fellows of Pembroke College Library in Cambridge, and the Trustees of the British Museum in London, for the permit to reproduce the manuscript drawings, shown in this volume.

F. D. P.

Introduction

Philo of Byzantium, also called Philo the Mechanic, has been famous, since ancient times, as a teacher of science and technology.

He may also be seen as an exponent of innovation. One of his surviving books, which deals with mechanized weapons, reports almost constantly on mechanical experiments, including experiments of his own, and on tests of new construction and of resulting improvements of performance. In his *Pneumatica* he begins with discoveries in fluidphysics, made and debated by natural philosophers; he adds extensive demonstrations of the discovered phenomena; and he gives a consistent, if not a final explanation, of the physical processes. His work has areas of similarity with modern experimental research, as well as it has contacts with the life of Greek seafarers and colonizers: it is supported by a spirit of seeking and conquering new domains.

In contrast to other ancient teachers of science, Philo makes little use of dialectics and mathematics. He rather shows the operation of devices which can lead common people to an intuitive understanding. With this method he had enormous success: several of the demonstrations described in his book are still given today, in schools of elementary physical science.

He conducts these demonstrations with devices developed from household tools, such as bottles and cups. His most sophisticated unit is the controlled vent: an antecedent of the modern pipette and an early substitute for a mechanized valve.

These beginnings are simple; what Philo developed is impressive. The *Mechanical System (Mēchanikē Syntaxis)*, completed by him about 200 B.C., may be considered "the first treatise of technology". Its *Pneumatica* has been called "The first experimental physics".[1] Earlier books of related kind were lost, while Philo's work influenced important successors.

Some successors called Philo a "Hermetic". According to a legend of the early middle ages, an Egyptian god and king, called Hermes Trismegistus, had founded magic science and had taught it to Pythagoras, from whom it sup-

[1] Sarton, II, p. 345; Hammer-Jensen, p. 414. NOTE ABOUT CITATIONS: for these and similar works, which are cited repeatedly in this edition, the Bibliography gives the complete names and titles. For other works, cited only once or a few times, complete names and titles are given in the first footnote citing them; in case of further citation, the number of that note is repeated after the brief name.

posedly descended to Philo and others.[2] Due to this legend, Philo became quite famous as an exponent of magic science. Near the end of the middle ages he, as the supposed inventor of certain unusual furnaces, was called "The Great Philo".[3]

At present, the understanding of Philo's works is unsettled. One modern historian believes that Philonian pneumatics led to the gradual recognition of "progress through experiments" Another writes that Philo already anticipated what was needed for the "Torricellian experiment", one of the foundations of modern physics.[4] Still others, by contrast, have heard so little about Philo that they, in their histories of fluid physics or of physics in general, hardly allude to him.[5]

Almost throughout the studies of such history, conducted thus far, the pertinent, pneumatic-archaeological evidence has been disregarded, and little attention has been given to the evidence of the authentic manuscript drawings. Such studies could hardly determine the pattern of progress and retrogression, or the changes of direction, in such history.

This edition of Philo's *Pneumatica* is intended to present such evidence, together with a fresh review of the texts, in order to provide a sound basis for historic conclusions, which the reader may draw.

---

[2] AlKindus, ed. Wiedemann, pp. 223–229.

[3] K. Kyeser, *Bellifortis*, ed. G. Quarg, Düsseldorf, 1967, fo. 112, b 3; Francesco di Giorgio Martini, *Trattati*, ed. G. Maltese, Milan, 1967, p. 169.

[4] Sarton, II, p. 349; F. Mason, *A History of the Sciences*, New York, 1962, p. 315.

[5] H. Rouse and S. Ince, *History of Hydraulics*, New York, 1963, p. 20; G. A. Tokaty, *Fluidmechanics*, Henley-on-Thames, 1971, pp. 8–33; A. Valdarnini, *Il metodo sperimentale*, Florence, 1920, pp. 10–40.

Philo's Backgrund, Work, Influence, and Reputation

I. The Background

Some of the pneumatic devices, shown in this volume, had been made, pictured and described before Philo's time. Use had long been made of plain suction tubes and of bent siphon tubes for transferring liquids. Early antiquity had developed a relatively complex, pneumatic machine: the bellows, which draws air through a valve into a leather bag, to discharge this air through another valve.[6] This machine is mentioned in Homer's *Iliad*: the artisan-god Hephaestus is described as operating a set of bellows to fan the flames on his hearth. There he supposedly produces fabulous automata. Parts of the story may have been developed, and interpolated in the older text, during Philo's time. Another legendary account attributes the invention of the bellows and of other machines to Anacharsis, a Scythian who lived near Byzantium, between the times of Homer and of Philo.[7]

Since the present notes are not intended to trace the history of science or technology in general, these notes will disregard many details of the early record. They will be brief even in dealing with some parts of modern pneumatics, disregarded by Philo, such as the bellows. They will be very brief indeed in dealing with hydraulic parallels, such as the early valve-controlled water pumps. We must only try to explore the background and the repercussions of Philo's pneumatic demonstrations.

In this attempt we must consider both the direct, explicit evidence and the plausible implications of the facts shown by the ancient record. This record includes ancient antecedents of the Philonian demonstrations, largely supplied by archaeology. It also includes both ancient and medieval developments of the Philonian demonstrations, attributable mainly to Hero, to Arabic intermediaries, and to Renaissance successors. One of the latter developments is illustrated by the frontispiece of this volume (Figure 1). Here we see a Philonian-Heronian, pneumatic-hydraulic automaton, in a medieval design reproduced in 1693. We shall have occasion, in notes which follow, to mention stylistic as well

---

[6] Suction tubes: Singer, I, p. 280. Siphons: Rouse-Ince[5], p. 6. Bellows: T. Ewbank, *A Descriptive and Historical Account of Hydraulic and Other Machines*, New York, 1841 (also 1870 etc.): pp. 205–207, 233–235.

[7] *Iliad*, XVIII, 373ff., 417ff., 470ff.; Pauly-Wissowa, I, col. 2017.

as technical features of the drawing, and to compare them with similar features of drawings more directly attributed to Philo, such as the drawings to Chapters 18 and 22 of the Arabic manuscript, published here.

Particularly advanced, in ancient development of a pneumatic sort, were the liquid-lifting "*clepsydra*" sieves. This kind of sieve was a vessel, closed all around except for sieve holes in its bottom and a vent hole in its top. Some forms of it are indicated in Figure 2. When liquid was in the vessel and was lifted with it, this liquid only fell from the vessel, through to open sieve, when the vent was open.

The modern rediscovery of these vessels began, shortly after the discovery of Pompeii and Herculaneum, with occasional studies by archaeologists and historians of engineering.[8] Thus far an adequate summary of the recovered clepsydras has been published only once, in a little-known article, the major illustrations of which are reproduced in the present Figure 2.[9]

Different variants of the liquid-lifting clepsydra sieve have been excavated. Most of them can be dated quite reliably by their decorations. It appears that their earliest forms (Figures 2a and b) were almost as old, by the time of Philo, as his works are by the present time. They were well known in Minoan and Mycenean times, and they continued to be used by ordinary householders, until very recent times. One variant, often made as a pot of clay, had its vent-hole in a hollow handle, which sometimes was stirrup-shaped and symmetrically secured to the top of the pot. Another form, often made of bronze, used a straight tube instead of this handle (Figure 2, c and b). Sometimes the vent-hole was provided in the integral cover (Figure 2a), and any handle then served only as a support.[10] Other variants added internal partitions and ducts, no doubt for the demonstration of surprising effects: Figure 2d. It seems that astonishing effects were also achieved without the use of a clepsydra "sieve" in the bottom of the vessel: Figure 2e and f.

---

[8] Ewbank[6], pp. 6f., 61–70, 94–96. Deonna, pp. 207–209, 218–232.

[9] Ewbank[6], pp. 193–196; E. Viollet-le-Duc, *Dictionnaire raisonné du mobilier*, Paris, 1871–75, II, pp. 16f.; E. Pottier, R. Zahn and others (c. 1880–1905), cited by Deonna; Daremberg-Saglio, s.v. "Siphon (VI., crible sphérique)"; M. Ebert, *Reallexicon der Vorgeschichte*, Berlin, 1924–32, XII, pp. 81f.

[10] Figures 2a and b (copied from Figures 8 and 9 of Deonna) show clepsydra sieves found in Mycenae and Pompeii. Vases of these types are now in Athens, Berlin, Boston, Crete, Naples etc. Also see Ewbank[6], pp. 193–200. Figure 2c shows a type found in Rhodes and Boeotia (and in western Europe: Ewbank, p. 194; Viollet-le-Duc[9]). Samples are in Athens, Berlin, New York, Paris, etc.; Deonna, p. 222f., shows a pedestal-mounted variant. The Greek surprise vessels, Fig. 2d, e and f, are Deonna's Figures 12, 6 and 15, respectively.

The Philo manuscripts illustrate many of these forms, in their various conceptions of Chapters 2, 6 and 11. They do *not* show the time-measuring clepsydrae or water-clocks, better known to antiquarians. These had a wide-open top instead of the fingersized vent, had a single orifice in the bottom, instead of the sieve, and were generally much larger.[11]

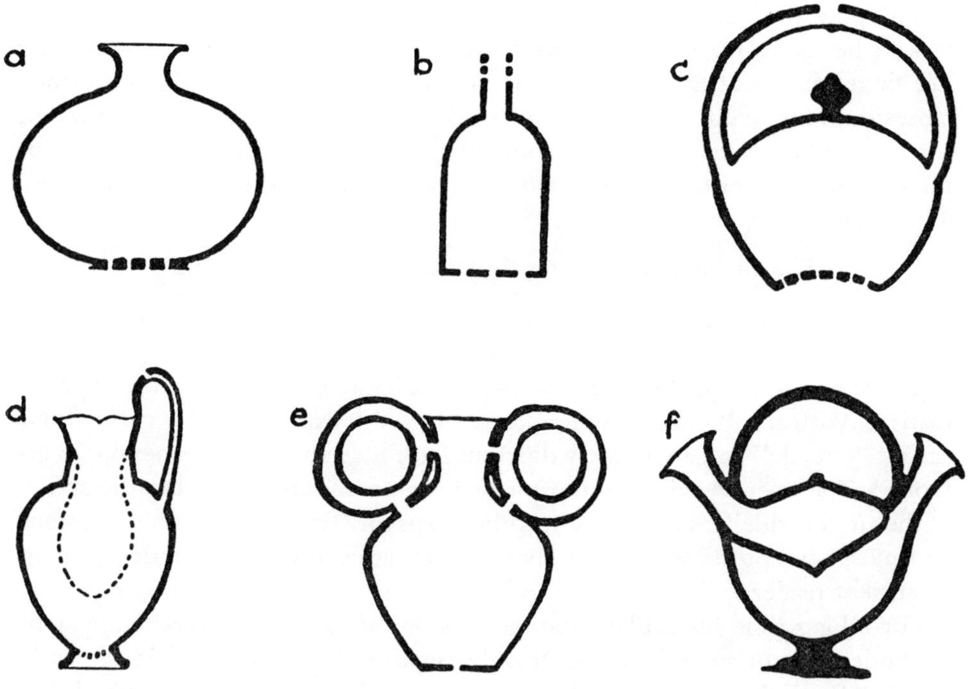

Figure 2. Ancient clepsydra sieves and surprise vessels.
From Greek excavations, reported in 1908

In antiquity, clepsydra sieves were widely used, thoroughly known, and incisively studied. An important reference to such a sieve occurs in a didactic poem *On Nature* (*Peri physeōs*), composed about 460 B.C. by Empedocles of Agrigentum. A large, pertinent fragment of the poem is preserved.[12] It contains the first clear reference to atmospheric pressure.

---

[11] Max C. P. Schmidt, *Die Entstehung der antiken Wasseruhr* (Part II of his *Kulturhistorische Beiträge*), Berlin, 1912, pp. 11–52, 65–80.

[12] Diels, *Fragmente*, Empedocles fragment 100.

The poet deals with respiration, and then tries to explain this biological process by comparing it with a known process in physics. A small girl, he says, may be seen playing with a clepsydra sieve. She submerges the vessel, as if to fill it, but first she has covered the vent with her finger. The vessel will not fill, that way: "none of the water enters." It is clear to the philosophical observer that "the bulk of air inside . . . keeps it out".

The poet then describes what happens when the girl has finally filled the vessel; he leaves it to the audience to realize that she does this by removing her finger from the exposed vent of the submerged vessel. When she then closes the vent again, and so lifts the filled vessel from the surrounding water, a strange phenomenon is seen: something opposes the natural tendency of the contained water to fall down from the open holes in the clepsydra bottom. The poet is clear about the reason:

> The air outside, striving to get in, holds the water in.[13]

This condition prevails, as he finally adds, until she re-opens the vent; there then follows a small rain from the openings of the sieve.

The action of the air "inside" the clepsydra became famous in ancient pneumatics; Aristotle, Philo and others refer to it very explicitly. The verse about the air "outside" had a strangely different fate: it seemed to be generally overlooked. The poet described and explained the procedure, involving the action of the air "outside", as clearly as the other clepsydra functions, but his reference to atmospheric pressure may have been too far advanced for the understanding of ancient readers.

For a long time his explanation was too far advanced to become apparent to modern historians of science. It only began to be understood in the early 1900's.[14] Since then the remarkable verse has acquired its deserved fame; it has rendered obsolete much that had been written about the history of pneumatics. A major problem of today, in a review of writings on pneumatics, is not how close Philo came to Torricelli, but how far he was from Empedocles.

---

[13] *αἰθὴρ δ'ἐκτὸς ἔσω λελιημένος ὄμβρον ἐρύκῃ.*

[14] This explanation was not noted, for example, by R. Caverni, *Storia del metodo sperimentale in Italia*, Florence, 1891–1900; by P. Duhem, *Le système du monde*, Paris, 1913–59; by Deonna; or by M. Schmidt[11]. It was noted vaguely (as referring to "weight of air") by E. Gerland, *Geschichte der Physik*, Munich, 1913, p. 37 and C. de Waard, *L'expérience barométrique*, Thouars, 1936, p. 20. Better accounts were given by J. Jeans, *The Growth of Physical Science*, Cambridge, 1948, p. 57, and by followers: see D. O'Brien, "Empedocles' Theories . . .," in *Journal of Hellenic Studies*, XC, 1970, pp. 140–179; J. Bollack, *Empédocle*, Paris, 1965–69, III, pp. 469–494.

Empedocles had not been alone in writing about the force of the outer air, impelling it to seek entrance into the clepsydra sieve and thereby to support the water inside the sieve. About the time of his poem there appeared another book On Nature, this one in prose, written by Anaxagoras. According to later restatements, his book dealt with the clepsydra sieve, and also with the workings of an air-inflated, weight-supporting "askos," (a wine-skin; a simple variant of the Homeric bellows). This wine-skin as well as the clepsydra-sieve could be seen in games played by children. Both devices also continued to be discussed by natural philosophers. For example, at least a confused recollection of the retention of water in the clepsydra was recorded near the end of the V. century B.C. by a Hippocratic writer, Aethlius.[15]

The greatest interest of philosophers, however, was concentrated on conceptual and metaphysical fundamentals: on space, matter, "the void," and the limits of the universe. Parmenides, shortly before Empedocles, had concerned himself with the world then known; he had totally denied the possibility and the existence of certain facts indicated by prior observation. He had, in fact, gone so far as to deny the possibility of "motion". He also had denied the possibility of a "void," perhaps in response to suggestions of Anaximander, who had written an earlier book On Nature, visualizing a cosmic void which contains a plurality or infinity of universes. The school of Parmenides is noted for the logical fallacies, half-seriously announced by it. Its view of the vacuum may be considered such a fallacy. This school identified void space as "that which is not," although by proper definition of course it only is "that which has no matter in it." With pseudo-simplicity, this school announced that "that which is not" is not.[16]

This obviously defective statement was strangely successful. For example, Empedocles is reported to have repeated: "Nothing is void." The opinion was unfounded. At best one might say that it seemed compatible with the observation of atmospheric pressure, so long as the physical condition of space outside the atmosphere had not been observed. One could still say that a continuum, not a void, is maintained when the pressure of the outer air, acting on the underside of the clepsydra, keeps the water inside the open sieve.

Plato, in the IV. century B.C., gave explicit expression to this latter opinion. He assumed a law of conservation of a *continuum* throughout the world. He further believed that such laws are founded on eternal, unchange-

---

15 Mentioned by de Waard[14], p. 64. About Anaxagoras: Diels, *Fragmente*, part 69 of his "Lehre;" D. O'Brien, "Anaxagoras and Empedocles," in *Journal of Hellenic Studies*, LXXXVIII, 1968, pp. 93–113. About the game of wine-skins: Daremberg-Saglio s.v. *askos, askolia.*

16 Sarton, I, pp. 173–175f., 245.

able, mathematically fixed "ideas." He therefore considered experimentation about physical laws unimportant. Although Plato's mind was most inventive, not only in a philosophical or poetical sense, (a late ancient writer credited him, for example, with the invention of the bell-ringing mechanism for the water-clock), yet there are authentic statements in his famous dialogues, which tend to ridicule the contemporary observers of physics and to deny the value of their work. In such statements the book of Anaxagoras is described as "available in the market, for no more than a drachma," but as containing "mere absurdities and ... no causality for the order of the world, except causes like air and aether and water."[17]

A major, physical observation had been made in Socrates' time, which Plato and his followers totally disregarded. Leucippus, followed by his pupil Democritus, had noted that "some things are observed to contract," even if they do not have visible, contractible pores. Detailed reports are not known, but of course the ancient observers could see that hot oil and hot metals contract, when chilled. Leucippus and Democritus proceeded to the question: into what do the materials contract? In reply, they recognized that inner voids must be dispersed in matter to allow this process to occur, and that the voids must be so small as to be invisible.[18] These observers also suggested that matter consists of equally small atoms, between which the voids are dispersed. They founded the vacuist and atomic theories.

Aside from being disregarded by Plato, these theories were resisted by his pupil and opponent, Aristotle (384–323 B.C.). The possibility of further development of the discoveries, announced by Empedocles, Anaxagoras and Democritus, was greatly impaired by this resistance. At least, however, Aristotle gave account of grounds for his views, in sharp contrast to the confusing attitudes of Parmenides and Plato.[19] It is only by the polemics of

---

[17] On the *continuum* of Empedocles: Diels, *Fragmente*, Empedocles fragments 13, 14; on that of Plato: *Timaeus*, 80C. On Platonic ideas: *Republic*, 596A to 597D. The references to Anaxagoras: *Apology*, 26D, E; *Phaedo*, 98B, C (cited after the translation of B. Jowett, Oxford, 1871). Also see Sarton, I, pp. 397–430; B. Russel, *History of Western Philosophy*, London, 1946, pp. 103–181. On the bell-ringing clockwork, attributed to Plato in late antiquity and with questionable nomenclature: M. Schmidt[11], p. 38.

[18] The quotation is from Aristotle, *Physica*, IV, 6, p. 213b 15. Also see C. Bailey, *The Greek Atomists*, Oxford, 1928, pp. 117–137; W. Heisenberg, *Physics and Philosophy*, New York, 1958, p. 66.

[19] Aristotle, *De respiratione*, 7, p. 473b 8 to 474a 5; *Physica*, IV, 6, p. 213a 23–34; *De coelo*, I, 10, p. 279b 7–14; *Meteorologica*, I, 3, p. 339b 29. Also see H.F. Cherniss, *Aristotle's Criticism of Presocratic Philosophy*, New York, 1964, pp. 164f.

Aristotle that we know the earlier statements about physical observation, dealing with clepsydra sieves and the like.

As is generally known, Aristotle agreed with his teacher in assuming a world centered on a terrestrial globe, while the two men disagreed abouth the motions of the cosmic bodies. Against Plato and with Democritus, Aristotle recognized the basic value of observation.[20] At the same time he practiced bold and often unfounded dialectics, which in some questions led him to views less defensible than those of Plato.

In pneumatics, Aristotle postulated an air which has "levity" and which thereby overcomes its observable weight or "gravity." Based on this unfounded and unclear postulate, he held that an outer sphere of meteoric "fire" is needed to keep the air in its "natural place."

Nevertheless, he was the first to explicitly recognize an experiment showing that air has weight. He also was the first to develop a somewhat detailed picture of the "*sphaira*" of "*atmis*." He described it as comprising an inner layer of cloud-bearing air and an outer layer of cloudless air, both layers being churned by storms due to solar heat. He was not sure whether the cloud-bearing air extends far beyond the highest mountain peaks, but he knew that "snow falls mostly on high mountains."[21]

Some errors of Aristotle came from the fact that the observations, available for an attempt to systematize them, were all too few. (The fall of heavy bodies is a well-known instance. Aristotle knew that specifically heavier bodies fall faster, in air, than specifically lighter ones. He had no quantitative equation between the specific weights and the falling velocities; no means to measure velocities at all; and therefore no means whatever to establish a valid equation. He postulated the equation. His postulate was very wrong, but almost two millennia were needed to dispose of the postulated equation and of the system based on it).

Some errors also came from dialectic distortion of observed facts. One of these distortions led to Aristotle's reaction to the ancient observation of atmospheric pressure. He assumed, very questionably, that Anaxagoras and "others" had attempted, by their study of clepsydra sieves and wine skins, to

---

[20] Aristotle, *De generatione*, I, 2, p. 316a 5–10. Also see *Metaphysica* I, 3, pp. 983b 6 to 988a 16 and its analysis by A.E. Taylor, *Aristotle on his Predecessors*, ed. Chicago, 1949, pp. 67–139, and L. Bourgey, *Observation et expérience chez Aristote*, Paris, 1955, pp. 78–82 and especially 83.

[21] *De coelo*, I, 3; IV, 5; pp. 269b 20–30; 313a 25; *Meteorologica*, I, 3; 9–14; pp. 339a 34 to 341a 10; 346b 15 to 353a 26.

"prove the impossibility of a void." Next, he stated, most erroneously, that these predecessors

> merely give an ingenious demonstration that air is something . . .[22]

The word "merely," in this opinion, is one of Aristotle's fatal errors. It overlooks his predecessors' observation and recognition of the pressure exerted by outer, unconfined air. As a result, his system proceeds to its labored, invalid dialectics about the non-existence of a vacuum.

Aristotle argues that something must be wrong with the view of the atomists and vacuists that only the presence of voids in matter can explain what happens when "things are observed to contract." Here he begins by rejecting the sophism of Parmenides and the Platonic view about the continuum. Then, however, proceeding along circuitous, dialectic paths, he repeats this sophism and this view in substance. He relies on a definition of „space," calling it "*The* boundary of *the* containing body, at which it is in contact with *the* contained body."[23] In the investigation of space, matter, and space without matter, this definition is a bad *petitio principii*, aside from its use of the undefined concepts "boundary" and „contact". His own logics damn the use of such a definition, in such an investigation. However, due to his logical blunder, his physical system became sharply anti-Democritian. Plato had assumed a law of preservation of a continuum between substances; Aristotle postulated a law of preservation of a continuum in and between substances.

Other Aristotelian remarks, including several on pneumatic observations, had greater physical merit. Even if they were not integrated into his system, but mentioned only marginally, they may have aided later pneumaticists. We will be faced with some such observations, in notes dealing with air weighing, air pumping, and air rarefying methods (discussed by Philo in his Chapters 2, 3, 6, 7 and 8, or at least alluded to in those chapters).

The problematic nature of Aristotle's doctrine was recognized by one of his early successors.

This successor was Strato, known as The Physicist. He flourished during several decades about the year 300 B.C., beginning in Athens, continuing in Alexandria, and ending again in Athens. There he became the head of the Aristotelian school. He was far from accepting all pronouncements of the Aristotelian system. His book about *The Void (De vacuo)*, according to rediscovered fragments, rejected the anti-vacuist part of the system.[24] Pertinent

---

22 *Physica*, IV, 6, p. 213a 24227; *De coelo*, II, 13, p. 294b 22–23.

23 *Physica*, IV, 2–9, pp. 209a 26 to 217b 28.

24 Diels, Straton, pp. 110–119; H.B. Gottschalk in *Proceedings*, *Leeds Philos. and Lit. Society*, Lit. and Hist. Section, XI, 1965, pp. 95–182.

details are unknown, but one fact is clear: after Strato and for the remaining centuries of the school's existence in Athens, the Aristotelians refrained from new attempts to systematize physical observations, and did not fully insist on the master's original system. Scientific writing became more monographic.

Such writing was known from famous, earlier examples. Military engineering books of the IV. century B.C. had offered accounts of tactics, horsemanship and the like. The *Tactical Commentary* of Aeneas Tacticus had described siege towers, chariots, catapults, clepsydra time clocks for use in camps, and a telegraph system.[25] It is unknown whether that book was illustrated, but there is evidence to indicate that it was. Surviving copies of somewhat later books have illustrations of remarkable uniformity. They suggest that Aeneas (if not Philo) had introduced a standard form or style of illustration.[26]

In pure mathematics, the *Elements* of Euclid, written in Strato's time, acquired unique fame and influence. In physics their methods were applied most rigorously by Archimedes (III–2 B.C.), so particularly with regard to the fundamental facts of hydrostatic equilibrium and buoyancy, which were basically or qualitatively known from Aristotle. As to the nature of fluids, Archimedes was reticent. It seems he remained closer to Aristotle, in this respect, than Strato did.[27] However, this may be only an appearance. It also seems that Archimedes was a friend of Aristarchus, one of the great opponents of the Aristotelian world-picture.

During the same time, the III. century B.C., a very different kind of monographic, pneumatic work was produced. It was the work of Ctesibius of Alexandria, which described major inventions and developments of the author. The book is lost, and only some of its contents are recorded; it is quite unknown how it related to the problems of Aristotelian or Stratonian theory, for example as to vacua. The record shows that Ctesibius developed and designed the famous machine which still is called "Ctesibian pump:" the basic form of a piston pump with check valves, similar to those of the prehistoric bellows. He also designed and built the "Ctesibian organ," wherein

[25] Ed. L.W. Hunter, Oxford, 1927, pp. 153, 219, 222f. Also see Max Jähns, *Geschichte der Kriegswissenschaften*, Munich, 1889, pp. 26, 29. About his time clocks: M. Schmidt[11], pp. 35f., 98. About this influence on Philo: Haase, pp. 430, 433. Some facts of early bibliography: Sarton, *Introduction*, I, p. 119.

[26] *Veteres mathematici*, pp. 7, 16, 27; Jähns[25], pp. 19–26.

[27] His hydrostatic treatise (Archimedes, *Opera omnia*, ed. J.-L. Heiberg, Leipzig, 1880, 1881, II, p. 354; Sarton, II, p. 77) postulates that a fluid transmits pressure by "parts lying evenly and being continuous."

such a pump acted on water to compress air into wind chest sections, controlled by keyboard-actuated valves; Ctesibius called it *hydraulis*. In addition he developed water clocks with new *parerga*; a "singing *cornucopia*;" and "systems of blackbirds singing by means of water works . . ., figures that drink and move, and other things that are found pleasing."[28]

Such systems were developed further by his pupils; they were most successful as educational-entertaining devices. In medieval copies of the ancient illustrations they are sometimes displayed on or between columns with modified Corinthian capitals (as were the column supports of war machines. An example may be seen in the automaton of Figure 1).[29]

In other illustrations and in archaeological samples, pneumatic devices are mounted on pedestals of their own. Many of these are in form of disk-shaped feet (Figure 2a) or of hollow cones and the like (Figure 2d). Later they were developed into rectangular or circular basins.

Ctesibius also developed the military machines of his time, continuing the work of Aeneas Tacticus. For example we read of Ctesibian experiments with resilient, metallic parts, substituted in lieu of the springs of wood or horn or hair strands, used in earlier bows and catapults. We also read of Ctesibian experiments with the resiliency of air compressed in a simplified kind of Ctesibian pump.[30] These were experiments of unique boldness, and of almost incalculable value for the expansion of physical and technical notions. Needless to say, they did not result in immediate adoption of the proposed devices, but they must have resulted in a general and most important re-assessment of material characteristics; no one had, previously, conceived of elasticity as a characteristic of metal or of air.

One may be tempted to assume that Ctesibius also engaged in public hydraulic works. There were other engineers who built such works for urban water supply, agricultural irrigation, and military operations. How-

---

[28] For the quotations: Vitruvius, IX, viii, 2; X, vi, viii; Athenaeus of Naucratis, XI, xxvii. Also see Ewbank[6], pp. 266–268; Carra de Vaux, Hydraulis, *passim ;* Drachmann, pp. 1–41; and C. Merckel, *Die Ingenieurtechnik im Altertum*, Berlin, 1899, pp. 37f. The "figures that drink" may imply a use of the bent siphon.

[29] See Figures 1, 22, 36–39. Philo, *Belopoiica*, p. 66 mentions the use of war machines with support posts having embellishing capitals.

[30] Philo, *Belopoiica*, pp. 77f. As noted by R.S. Woodbury, *History of the Lathe*, Boston, 1961, p. 30, grinding and lapping procedures must have been used for the production of cylinders and pistons on the metal-working lathe of Ctesibian and Philonian times. Drachmann, p. 50, believed that such procedures began only in the Arabic middle ages.

ever there is no ancient record which indicates that Ctesibius, or his direct pupils, engaged in this further type of activity.[31]

Regarding the personality of Ctesibius, there is an ancient story about his modest beginnings, as son of a barber. According to the story he showed an early talent for the construction of mechanical devices; at one time he wished to hang a barber's mirror so that it could be raised and lowered by a concealed cord; he drew this cord, with a leaden counterweight at one end, through a pipe; when the weight descended into a restricted part of the pipe, a sound was heard, as air became compressed. This finding, it is said, animated the boy to pneumatic studies, and in due course it led to his inventions in this field. He became highly honored at the royal court, but continued to live in a modest quarter of Alexandria. His wife gave musical performances on the *hydraulis* or Ctesibian organ.[32]

II. Philo's Work[33]

Philo often speaks of Ctesibius, always in connection with experiments. It is not clear during what period of time he knew the older master; some Ctesibian works were only reported to him by intermediaries. Nor is it clear at what time Philo came from his birthplace, Byzantium, to Alexandria. He also spent time in Rhodes, which is on the way between the two cities. The three places were colonies, founded by men from the Greek mainland. They were active centers of trade.[34]

Alexandria was the newest and most rapidly developing colony known to Philo. Her story may have contributed to his interest in innovation. She had been founded very shortly before his time, and she had developed, within a few decades, into a teeming metropolis. The Greeks of earlier centers and colonies considered themselves more sophisticated than the Macedonian

---

[31] See below, Chapters 5, 61, 64, 65. Rostovtzeff, pp. 256, 360, 1073, 1232 describes such works of Cleon and Theodorus, contemporaries of Philo. Also see Merkkel[28], *passim*; M. S. Drower in Singer, I, pp. 520–536, and R. J. Forbes in Singer, II, pp. 663–694.

[32] Athenaeus[28], pp. 203–209; Drachmann, Second Ktesibios, p. 4.

[33] The following relies largely on Haase, pp. 428ff.; Hammer-Jensen, pp. 414–423, 480–497; and Drachmann, pp. 41–44.

[34] On Alexandria: Rostovtzeff, pp. 156–158, 300–316, 417; on Byzantium and Rhodes: *ibid.*, pp. 589–591, 676–691. (The account of W. W. Tarn, *Hellenistic Civilisation*, London, 1966, pp. 177–207, is based on obsolete sources, and is lacking in the brilliant illustrations, provided by Rostovtzeff.)

masters of Alexandria,[35] but they admitted her leading position as a center of the sciences and the arts.

Many scientists lived at the Alexandrian "*Museum*," and it is probable that Philo lived there. He called Alexandria a place where

> builders have recently been provided with large public assistance, by kings who love science and the arts.[36]

The Museum, with its famous library filling a separate building, was part of a royal compound, overlooking the city's "outer harbor" and its tall *Pharos*. The scientists met daily for a common meal, where one of them presided as a royal official. At least from the third to the first century, B.C. such men spent their days, doing research, with leisure from economic worries.[37] It seems that Philo enjoyed such leisure.

Those of his books which are preserved in Greek are technically, if not stylistically, very polished. Although they deal with matters of novel, multi-faceted nature, they do this with complete mastery of contemporary science and technology. To achieve such writing calls for time as well as talent. Years may have gone into the planning, correcting, and final formulating of those books. The preparations for the underlying experiments must have consumed equal time.

Archaeological artifacts, made by or for Philo, have not been identified. However, clepsydra sieves and surprise vessels of his and his teacher's time, identified by their surface finish, are known in large numbers, and an occasional one, among them, bears the potter's signature.[38] We may assume that Philo cooperated with artisans of Alexandria, who were proud individuals in their own right.

We do not know his exact dates, and do not even know how long he lived. Linguistic analysis shows that he wrote the kind of Greek that was used about 200 B.C.[39] Unfortunately he does not mention a large number of datable events, the last of which would indicate the time when he wrote his book.

---

[35] Theocritus, XIV.

[36] Belopoiica, p. 50. I cannot agree with Krafft, pp. 348–350, who interprets Philo's statement as an "envious" comparison with a *former* condition. The interpretation is based on the pessimistic evaluation of Philo's time (the time of Ptolemaeus III. and IV.), suggested by Polybius. More reliable is the account of Rostovtzeff, pp. 705–736.

[37] Rostovtzeff, pp. 1084f.; Mason[4], pp. 49f.; Sarton, II, pp. 29–31, 141–155.

[38] Deonna, pp. 223f.: a VI. century B.C. vase of the type of Figure 2d, signed by the potter Kolchos.

[39] M. Arnim, *De Philonis . . . dicendi genere*, Greifswald, 1912, p. 164.

One of the few political events of known date, mentioned by him, is the defense of the Rhodians against Demetrius Poliorcetes, king of Macedonia. This event had taken place some decades before Philo was born.

Parts of Philo's *Mechanical System* are lost, but there is internal evidence for its original structure.

It began with an *Introductory Book, Eisagogē*, which dealt with the mathematics then known and needed by workers in "mechanical" physics and technology. This book may have dealt also with such introductory matters as the mechanics of rope wheels and gear wheels, previously or contemporaneously discussed in a booklet of the Aristotelian school which was concerned with mechanical problems. A few pages of mathematics, in Greek, are preserved from Philo's introductory book. They show that he was conversant with Euclidian geometry, and in fact was expert in it; he occasionally devised a new and more powerful proof for a proposition.[40] According to Philo's work on pneumatics, his system also included a book or chapter on atoms and distributed vacua. This, along with other physics, may have been a non-mathematical part of the introductory book.[41]

There followed a book on *Levers (Mochlica)*, perhaps concerned with contemporary mechanics more generally. Then came a book of applied science, dealing with *Harbors (Limenopoiica)*. These two books are generally assumed to be among the lost works.[42] The then following books, on the other hand, are preserved, in Greek; they deal with war engines, their construction, improvement, and use. One of these is called *Belopoiica*; two others are *Parasceuastica*, defense tactics, and *Poliorcetica*, siege tactics. According to a remark, inserted shortly before the end of the war engine book, he then intended to begin the work on pneumatics.[43]

---

[40] For the *Mechanical Problems ("On the Lever")*: Drachmann, *Technology*, pp. 10, 13–18. For Philo's proof of Euclid I, 8, reported by Proclus (V–1): Heath, ed. of Euclid's *Elements*, New York, 1965, I, pp. 263f. For Philo's analysis of Euclid V, def. 9 (duplication of cube), reported by Eutocius (VI–1): Heiberg[27], III, pp. 72–75. Generally about Philo's "mechanics, pneumatics and hydraulics", as cited by Pappus and Theon of Alexandria (III–1 and IV–2) and Tzetzes of Byzantium (XII–2): F. Hultsch, introduction and notes to vol. III of *Pappi Alexandrini Collectionis quae supersunt*, Berlin 1878, pp. VIII f., 1015, 1022.

[41] See Chapter 3 of the *Pneumatica*, below; Haase, p. 430.

[42] But, as noted hereafter, there are remarks of Kircher dealing with mechanical devices described in a Philonian book "*On the Five Static Powers.*"

[43] *Belopoiica*, p. 77; Haase, p. 432. "He [Ctesibius] knew from the reading of pneumatic teachings [of Strato?], which will also be treated by us . . ."

It is unknown how accurate and how complete the internal cross-references of Philo's work are, but they have been confirmed by the discovery of the *Pneumatica*. Additional Philonian works have also been reported; so for example a book *On the Five Static Powers (De quinque facultatibus staticis)*, attributed to Philo in 1664, and a book called *Automatopoiica*, reported in late Hellenistic times.[44]

Perhaps we may conclude that the work, after its basic introduction, presented the science of mechanics in the book of static powers; continued with applied mechanics in the books on harbors, levers (harbor cranes), war engines, and automata, and then presented theoretical pneumatics and such applications as were then conceivable.

The illustrations of Philo's book on automata related mainly, if not exclusively, to mechanized theater performances. It seems that the drawings showed his stage as a small model of a temple, standing on a tall pedestal. Columns flanked the stage. According to later illustrations—for example, Figure 1 in this edition—the columns had modified Corinthian capitals; these are also found in illustrations of war engines, as already mentioned. Either dramatic or scientific demonstrations were given on a stage between such columns. Philo's demonstrations were successful in Hellenistic times. His chief successor, Hero, wrote some centuries later: "... we have found nothing better among our predecessors ... than the notations of Philo of Byzantium." He added that he had "reviewed many copies" when confronted with apparent errors as to some details.[45]

Philo's style of writing may be reconstructed, to some extent, from the reports and quotations of Hero, and from the abbreviations of Philo's own works, surviving in Greek. The descriptions have much in common with works of Aeneas Tacticus, the predecessor in the field of war engines.[46] They use methods parallel to those of Euclid, but in ways different from the rigorous, mathematical presentations of Archimedes. Philo seems to have been capable of such presentation, but to have used it only in real mathematics, not in discussions or applications of physics.

The pneumatic demonstrations of Philo showed, in substance, how liquids can be lifted and then lowered, in siphons, and how they can be held in raised position, in clepsydra sieves and the like, until gravity is allowed to bring them down. This is the substance of the present Chapters 1 to 23. Beginning at or near Chapter 13, and continuing—with interruptions—to the vicinity

---

[44] Hero, I, pp. 405–411 (in Hero's *Automatopoiica*).
[45] Hero, I, pp. 412–453 *(ibid.)*.
[46] Jähns[25], pp. 26–29; also see A.R. Hall in Singer, II, pp. 208–213.

of Chapter 39, surprise and entertainment becomes an element of these demonstrations. No doubt this element is borrowed, to a large extent, from the work of Ctesibius. No attempt is made to suggest or discuss any play of alleged, magic powers. Everything is fully explained by rational physics. This rational character of the work also pervades the remainder of the surviving manuscripts, mainly Chapters 40 to 65, which relate to hydraulic, mechanical, and pyrotechnical rather than pneumatic demonstrations.

The demonstrations now offered by the Philonian *Pneumatica* are very effective, insofar as they are truly pneumatic; less so insofar as they are hydraulic, mechanical and pyrotechnic. It seems possible that either the mental powers of Philo diminished, when he reached those parts, or that his work was altered, proliferated and misarranged by followers endowed with lesser stylistic and educational powers. By way of documentary notes, presented as a second part of this Introduction, I will try to show the evidence bearing on this problem. At the present moment I will limit myself to outlining the nature and history of Philo's work in real pneumatics.

As an explanation of the demonstrated, pneumatic phenomena, he offered a theory which in some respects differed from all documented theories of former contributors, such as Empedocles, Anaxagoras and Strato, and which more radically differed from the dialectics of Parmenides, Plato and Aristotle: Philo's theory of cohesion. This theory, outlined in his Chapter 3, can be understood as postulating, with Democritus, that matter consists of particles, mixed with distributed voids. It adds that the particles of a substance cohere with one another, and with those of a contacting substance, by intermeshing in such voids. For further details, the text refers to a separate book or chapter, now lost. Erroneously, Philo believed that the forces of cohesion between the particles are sufficient to explain the phenomena of liquid-lifting siphons and clepsydras.[47]

The text also adds that no extended void—no void larger than the invisible ones between atoms—will be found in nature. Perhaps this fact was derived from the non-vacuist doctrines of Plato and Aristotle; Philo's text does not derive it from the dialectics of those philosophers. The text of the fuller discussion of particles and voids, as mentioned, is not preserved.

Such a discussion may have been part of Philo's *Introductory Book*, as already mentioned. It may also have been a chapter of the *Pneumatica*, or a chapter connected with it. If it was comparable, in length, to the purely pneumatic chapters presently known, it completed a book similar in length to Philo's book on war engines, which in turn (before abbreviation) had the approxi-

[47] A different interpretation of Philo's doctrine is given by Krafft, p. 368.

mate standard length of an ancient book. This standard length was set by the dimensions of a book roll. It is illustrated by the length of one of the *Gospels*; that is, it contains some 12,000 to 24,000 words.[48]

It also seems possible that Philo's original book on pneumatics discussed the elastic powers of air, which are disregarded in the existing *Pneumatica*, and are barely mentioned in a remark at the end of the *Belopoiica*. However, it may have been beyond the experimental and conceptual resources of Philo to explain such phenomena as the compression and rarefaction of air and the elastic return to normal, atmospheric pressure. As mentioned at various points in these notes, the concept of atmospheric pressure was within reach, but we do not know whether it was discussed by Philo.

It is also possible that he intentionally refrained from completing his work in this respect. He knew that it could be improved, but not perfected for all times. He wrote that his teacher had achieved improvement, "partly by learning from the earlier constructors, and partly by observation of later trials."[49] It was clear that he, Philo, had similarly achieved further improvement, and that work remained to be done.

Philo was concerned about competing practitioners, who claimed an inventor's fame, based on mere tinkering. He called it "improving and rearranging what has once been created."[50] He had run into such a man in the field of hydraulic works, according to a bitter remark in Chapter 5 of the Pneumatica. The man's identity and fate are unknown.

We may call it a major achievement of Strato, Ctesibius and Philo that they began systematic, visible, rational demonstrations of facts in the field of physics. These demonstrations, including those of entertaining kind, contributed strongly to the much-needed disproof of the anti-vacuist, anti-atomic and ultimately anti-observational dialectics of "classic" Greece. In particular these demonstrations were fit to overthrow the Platonic doctrine of unchangeable ideas. They showed to both citizens and philosophers of Alexandria that ideas were created and changed, by Strato, Ctesibius and Philo. For the progress of observational science these demonstrations were epoch-making. In the history of pneumatics, they were fundamental.

---

[48] For this standard see F. G. Kenyon, *Books and Readers in Ancient Greece and Rome*, Oxford, 1951, p. 64. The abbreviated *Belopoiica* text has about one-half this standard length; that of the Latin *Pneumatica*, about one quarter of it. Also see Rostovtzeff, p. 312 n.

[49] *Belopoiica*, p. 64f.

[50] *Belopoiica*, p. 58.

III. Philo's Influence in Late Antiquity and the Early Middle Ages

The books of Philo, individually sent to his friend Ariston, circulated in the ancient world; so particularly the military books, the topic of which was always of interest. However, the books circulated in small numbers, and not as complete sets of the *Mechanical System*.[51] In non-authentic ways, readers or editors probably combined pneumatic, mechanical and other chapters into compilations, beginning the development of medieval archetypes and possibly mixing Philonian chapters with work of their own or of others. They also abbreviated the authentic, Philonian texts, and no doubt altered at least details of the authentic drawings.

Some of the pneumatic concepts of Philo were also discussed in an anonymous work, called *Natural Problems*, which is usually, plausibly attributed to Aristotelians of Hellenistic times.[52] The writing of this book may have begun some decades before Philo's work, and finished long after Philo's work; thus it is possible that there were cross-influences.

The pneumatic chapter of these *Problems* began by describing simple beaker-submersion tests, which it attributed, plausibly, to Anaxagoras. The chapter continued with tests of clepsydra sieves. The authors pointed out that, aside from the requirements taught by Anaxagoras, the clepsydra sieve or beaker must be immersed vertically. If it is immersed obliquely, air escapes from its underside, and is not properly compressed; water then enters, and the simple, basic proof that "air is a body" no longer succeeds. In Philo's Chapter 2 the same thing was said, after a reference to unnamed "philosophers," who may well be identical with Anaxagoras and his contemporary, Empedocles; the books no doubt were preserved in the Peripatetic and Alexandrine libraries, and the demonstrations and observations, also including determinations not recorded in the basic books, were being continued.

Perhaps the most remarkable statement of the *Problems*, in the same chapter, dealt with the ultimate falling of the water from the clepsydra sieve. The water was said to fall by "forcing its way through the air outside." The statement seems to imply that the pressure of this outer air still was remembered, when Philo wrote his Chapter 11 and when his pupils read it.

---

[51] Haase, pp. 431, 434; also Sarton, p. 178 n. The main successor to Philo as military engineer, in Roman imperial times, was Apollodorus of Damascus and Rome. His *Poliorcetica (Veteres mathematici*, pp. 15–42*)* became the starting point of Byzantine and western developments: Jähns[25], pp. 101f.

[52] *Natural Problems*, XVI, 8, pp. 914b 9 to 915a 25 (paginated as part of Aristotle's works. About the actual time: Hammer-Jensen, p. 417).

It seems quite probable, under these circumstances, that there was discussion and even "considerable discussion" of the various pneumatic theories and observations, in the times of Philo and the next few generations. However, no details are known. It only appears that a bath construction, attributable to Philo, was developed by Sergius Orata, a Roman inventor of the next century, while military designs were developed for example by Athenaeus Poliorcetes of Cyzicus (near Byzantium) and by his follower Apollodorus of Damascus and Rome.

Attempts were also made on a larger scope to convey Greek science to Rome: so mainly by Vitruvius (I-2 B.C.), whose *Ten Books of Architecture* consider Ctesibius in some detail and Philo more generally, paraphrasing statements of Philo which dealt with the famous siege of Rhodes and with the problem of scale models for catapults.[53]

By the time of this Vitruvian paraphrase, copies of Philonian books must have reached Rome. From then on, Latin translations were eminently possible. Were such translations actually made in antiquity? The answer is unknown, but it is interesting to note that some existing, medieval copies of a Latin translation are closely combined with a booklet derived from Vitruvius. This booklet will be discussed hereafter, as one of several Gothic fragments of Philo's text. Some copies of Philo's *Pneumatica* must have circulated without Philo's name on them; they were later attributed to Aristotle.

In any event, Philo's pneumatic theory was familiar to educated Greeks. The author's namesake, Philo Judaeus (A.D. I-1), described air as an element endowed with "pulling power:"[54] an expression which was foreign to the works of classic authors; it reflected the theory of "cohesion" between air and water, taught in Philo's pneumatics.

Educated Romans may have been less aware of Philo. He is hardly mentioned in the *Natural History* of Pliny (A.D. I-2), or in the military books of imperial times.[55] It is, however, certain that his work was known to the great

---

[53] Athenaeus, *Peri mēchanēmatōn*, ed. *Veteres mathematici*, pp. 1–12; also see Jähns[25], p. 43. Vitruvius, on Philo in general, VII, Introduction, 14; on Rhodes and on scale models (Philo, *Belopoiica*, p. 55): Vitruvius, X, xvi, 3–8. About "considerable discussion:" Hammer-Jensen, p. 427.

[54] Philo Judaeus, "De eo quod a deo mittantur somnia," in his *Quae extant opera*, Paris, 1640, p. 568. Also see S. Sambursky, *The Physical World of Late Antiquity*, London, 1962, *passim*, about the Jewish Philo's views of other aspects of the physical world.

[55] Pliny, VII, 125 cites Ctesibius together with Philo of Athens, also known as Philo of Phalerum, an architect (IV-2 B.C.). About Pliny and the chief engineer of his time, Apollodorus: W.H. Stahl, *Roman Science*, New York, 1952, pp. 3–5, 101–119.

experimental physicist of late antiquity: Hero of Alexandria. He probably lived in Pliny's time or soon thereafter.[56]

In Hero's *Pneumatica*, the physics of Strato and Ctesibius and the general theory of Philo had clear echoes; the theory of cohesion had none. According to Hero, matter consists of particles mixed with distributed vacua, and these vacua can be enlarged or reduced, by applications of forces, although they naturally, elastically resist such deformation.[57] At a crucial point, Hero asks whether it is the impossibility of a vacuum [as asserted by Aristotle], which causes the resistance to a loss of pneumatic suction; his early, Greek manuscript suggests the answer "No." Unfortunately, his question and answer have been corrupted by modern editors, who have printed his text with anachronistic modernization of its punctuation.[58]

Hero repeats, and at some points independently develops the technology of Ctesibius and Philo, throughout his own *Pneumatica* and in certain parts of his *Automatopoiica*. In many constructions, Hero achieves improved effects by combining mechanical contrivances with the more purely pneumatic arrangements of Philo and his teacher. In particular, Hero makes use of rotary valves, floats, levers and mechanized valve actuators. These had only begun to appear, in elementary forms, in work attributable to Philo; they are used with greater effect in Hero's demonstrations. A splendid example of a rotary valve unit and mechanized actuator may be seen in the pneumatic automaton shown in the frontispiece of this book. In it, Hero develops one of the "vessels from which a variety of liquids may be made to flow." Much more modest variants of such vessels are shown, for example in Chapter 22 attributed to Philo. The book of Hero also shows new forms of piston pumps, organs, and other devices: new and ingenious combinations of older elements, just as the earlier, famous works had shown new combinations.

[56] So mainly Hammer-Jensen, Heronische Frage, pp. 43f. Drachmann, pp. 74f., and same, *Technology*, pp. 9f., insists on the first century A.D., in view of a Heronian reference to an eclipse of the sun, visible in Alexandria and Rome in A.D. 62. However, Hero may have written his reference to it long after the event.

[57] Hero, pp. 16ff. (Introduction and I, 2). Krafft, pp. 369f. Many details are in need of review.

[58] Hero (Greek Manuscript 519, Biblioteca Marciana, Venice, fo. 166r) wrote: "How does it come about that the water in the siphon, lifted up, does not fall down . . .? Because a continuous vacuum cannot exist? *In fact*, . . . unless the water is disrupted, by air entering through a hole drilled in the siphon, the water, resting on the underlying air, seeks to displace this air, but *the air*, *unable to yield*, *prevents the water from issuing*." The standard printed edition (*Opera*, Chapter I, 2: p. 36, lines 8–18) replaces the emphasized question mark (a raised point) and the term "*In fact*..." by a period, thereby totally corrupting Hero's statement.

In centuries after the Alexandrian masters, their works remained popular. Handbooks—quite unaffected by the metaphysical debates of the times—were copied and recopied to preserve the traditions of agriculture, dye-works, and miniature-painting, as those of surgeons and surveyors; some of these have been published in modern times. Similar handbooks preserved traditions of architecture, carpentry, masonry, and well-digging. As shown by the Philonian documents, some people also wanted to read about the pneumatic demonstrations. For such purposes, Greek archetypes were then developed, the contents of which are now reflected by copies of later translations, possibly corrupted by various errors.[59]

Some errors came into the record due to a "wave of Babylonian astrology" and a "fascination of the occult." In other words, late antiquity brought "Dark Ages" for the observational sciences; it brought real life mainly to Hermetic and Neo-Platonic pseudo-sciences. Pneumatics came to a standstill, and began to retrogress.[60]

At the same time some other sciences and techniques continued to progress. Artisans at centers of Babylonian, Chaldaean and Sassanian culture—collectively designated here as "Persian" artisans—contributed to the development of metal-working techniques, and stylistic characteristics. They improved the construction, if not the understanding of pneumatic devices. For example such artisans developed better valves and valve systems, along with new shapes and arrangements of vessels. Such, at least, is the impression that arises from certain chapters of the somewhat later, Arabic traditions, which are full of terms in Aramaic, Pahlavi and related languages, collectively designated here as "Persian."

The books of Greek scientists, written in the latest centuries of antiquity, were mathematically oriented. They tended to consider Philo and Hero as mathematicians, rather than a physicists or as teachers or inventors of technology. Such citations can be found in the writings of Pappus of Alexandria (III-2 A.D.), Proclus of Byzantium (V-2) and Eutocius of Ascalon (VI-2). The books on pneumatics were still copied or abbreviated, and perhaps were still used in the teaching of medicine and the like, but were no longer developed by new, progressive, physical thought, as they had been during the centuries from Empedocles to Hero. In the end the antivacuist

[59] Schmidt on Carra de Vaux, pp. 327f.; similarly Hultsch[40], p. 1015 n.

[60] The citations are from Mason[4], p. 49 and E. J. Dijksterhuis, *The Mechanization of the World Picture*, Oxford, 1961, pp. 45–49. About the Hermetics: Thorndike, II, pp. 214–229, 649–687; Sambursky[54], pp. 80f. Also see J. E. Sandys, *A History of Classic Scholarship*, ed. New York, 1964, I, pp. 664f. About early medieval Persian technology: P. Sykes, *A History of Persia*, London, 1963, I, pp. 380–488 *passim*.

dialectics of Aristotle prevailed over the vacuist demonstrations of the Alexandrians. The regression of science even led back to the animistic doctrine that "nature" has a "horror" of the void.[61]

Near the middle of the VI. century, Emperor Justinian closed the ancient academies, which had been the chief repositories of scientific records. Some scientists of Athens then sought refuge at the court of "Persia," perhaps preceded and followed by Greek artisans, skilled in the construction of mechanical and pneumatic devices. The scientists later turned to the old Museum of Alexandria. Commentaries to Aristotelian books, written under these sad conditions, could hardly do more than preserve a few fragments of the former, observational physics.[62]

In the next century, the *Etymologies* of bishop Isidor of Seville were considered a fountain of scientific information. In actual fact, the work shows a most regrettable lack of information on every Greek and Hellenistic teaching that had tried to explain the problems of pneumatics. It speaks of elementary concepts in naive, barbaric terms, for example: "Air is emptiness, with mostly rarity admixed to it."[63] Ancient words had survived; their use showed abysmal ignorance of the ancient concepts.

The Byzantines, as speakers and readers of Greek, were somewhat better informed even in the dark ages. A man called Philo Sophista Byzantius wrote pertinently, if not significantly, about the "Seven Wonders" of the ancient world. The so-called Pseudo-Hero or Hero Byzantius rewrote the books of Hero with detail improvements. In the west, one Landfridus rewrote the pneumatics of Hero or Pseudo-Hero, and included practical, Vitruvian rules for well-builders and mill-constructors, in a book *De vacuo et inani*.[64] Byzantine chronicles, including a late one by Tzetzes (A.D. XII-2), showed that the people still had some idea of the work done by Philo and Hero.[65]

---

61 For the mathematical writers: Hultsch[40], Heath[40], Heiberg[40], *loc. cit.* For the history of the *horror vacui*: Duhem[14], VIII, pp. 16ff., 126ff. For the "vacuum" in physiology: Galen, ed. K.G. Kühn, Leipzig, 1821–1833, IV, p. 222.

62 Sandys[60], I, pp. 85–87, 375; Diels, Straton, pp. 112–116.

63 Isidorus, *Etymologiarum sive Originum libri XX*, ed. A.P. Migne, Paris, 1850, *Patrologia Latina*, LXXXII, xiii, De mundo et partibus, pp. 471f. Also see Sambursky[54], pp. 80f.; A.D. White, *A History of the Warfare of Science with Theology*, New York, 1896, p. 326 (reprinted 1960, New York).

64 Duhem[14], VIII, pp. 121, 128f. Also see E. Grant in *Speculum*, XLVI, 1971, p. 669 n.

65 On Philo Sophista: Sarton, II, pp. 346f.; Pauly-Wissowa, VIII, i, *s.v.* Philon. On Hero Byzantinus: Drachmann, pp. 161–167. On the Tzetzes chronicles (XII-2 A.D.): Hultsch[40], p. VIII.

East of Byzantium, the time between Hero and Pseudo-Hero gave rise to a "Persian" tradition, based on Philo's pneumatics.[66] During the same time, another variant Philo's work may have been produced in the northwest of Europe. It is indicated by several pieces of evidence, surviving from somewhat later times. This northwestern variant seems to have used peculiar elements of graphic style, which reappear in later (eastern) illustrations. They include the strange forms of column capitals, already mentioned (Figures 1 and 22), which are best known in "Carolingian" art. They also include equally strange forms of human and animal figures (Figures 26 to 39), which have close parallels in the "Celtic" art of the VII. century, also called Irish art or Anglo-Saxon art. These elements can hardly be explained if not by assuming that a version of Philo's work was produced in one of the northwestern, monastic centers of Celtic art; that this happend shortly before or during Carolingian times; and that this version in turn influenced a later artist.[67] As Philo had been known in the west since the time of Vitruvius, this evidence for a Celtic variant is not particularly surprising, although the variant itself, like the Persian one, is lost.

More definite facts are recorded about the Arabic scientific effort. This began in the IX. century, about a hundred years after the conquest of Persia by Arabic armies. A first contribution to the development of pneumatics is attributed to the "Sons of Moses," Banu Musa. Their father, Musa Ibn Shakir, had been a famous robber; he "repented" and became an equally famous astronomer and court astrologer. The sons were educated, after his death, by another court astrologer. Two of these brothers took an interest in mathematics, astronomy, and astrological medicine; the third of them, Ahmed, wrote a *Work on Ingenious (Pneumatic) Devices*. The brothers became quite rich. They sent expeditions to Constantinople to purchase Greek works of science. At home in Baghdad they had some such works translated by the most recognized Arabic interpreters. At the same time the brothers engaged in scientific work of their own. Their activity initiated the "Hellenistic Renewal at Baghdad."[68]

---

[66] Carra de Vaux, pp. 39f., 230–235; his Advance Notice, p. 29; Hauser, introduction to his edition of Ibn Musa, p. 31.

[67] For other cases of such influence, see A. Boeckler, *Abendländische Miniaturen bis zum Ausgang der romanischen Zeit*, Berlin, 1930, Pl. 6, 1, 9, 11; 39, 55, 63, 64, 81 etc. For western, medieval influence on eastern books: Hauser on Ibn Musa, p. 27; Wiedemann, Beitrag VI in vol. XXXVIII, 1906, p. 23; K. Weitzmann, *Die byzantinische Buchmalerei des IX. und X. Jahrhunderts*, Berlin, 1935, pp. 67, 75f. Also see Duhem[14], VIII, pp. 24f.: an Arabic citation of Roger Bacon as "Bakun."

[68] Hauser on Ibn Musa, p. 7; Wiedemann, Beitrag XII in vol. XXXIX, 1907, pp. 201–204, on lamps of Ibn Musa.

The work of Ahmed Ibn Musa is preserved in copies of the XII. century. These have been published, with great care, in German. Only a minor part of them reflects ideas from the pneumatics books of Philo and Hero, in form of selected extracts or fragments. A larger number of chapters and drawings, in the Arabic work, are freely developed variants; they include some technical refinements. We will have occasion (in notes to Philo's Chapters 10, 13, 17, 20 and so on down to 60) to compare concepts of the *Pneumatica* copies, attributed to Philo, with the relatively close versions of Ahmed. The additional, more independent, Arabic chapters of Ahmed's work show the use of float valves along with(of the ancient "flooded vents"), and other devices which are of relative, technical sophistication. At the same time, the entire text of Ahmed is conceived in extremely naive terms. It hardly seems that he had access to the Greek text of Philo, which is reflected, beginning five centuries later, in western and eastern manuscripts.

Somewhat later in the IX. century, one of several writers called AlKindus—this one in Egypt—wrote a book on *Automata and Wondrous Machines*, Liber AlKindi *De automatis et thaumantiis*. This work, by itself, is lost, but there are Latin excerpts from it in a Venetian notebook of the XV. century, and there is a strange, indirect reference to it in the XVII. century, suggesting that the Latin Philo tradition was derived from AlKindus' book. The suggestion is questionable, in view of other evidence and documentary details. According to the Latin excerpts, the Arabic work contained variants, not copies, of the descriptions attributable to Philo and Ibn Musa. In another work, AlKindus designated the physicists of ancient Greece as pupils of the Egyptian god-king Hermes. He also said that Philo was "learned in water-wheels and mills and in motions performed by ingenious devices."[69] This latter expression may seem interesting, as references to Philonian water wheels and mills also occur in a later, Arabic book of "extracts." However, little or no ancient authority exists for water wheels built or designed or studied by Philo. The Egyptian's story about such devices may be as fantastic as the legend of "Hermes"."

Such fantasies were widespread. Only names of recent authors were remembered with some accuracy. For example AlKindus of Egypt identified himself as an admiror of the Banu Musa, in contrast to his name-sake in

---

69 AlKindus, ed. Wiedemann, pp. 230, 233, B. d'Herbelot, *Bibliothèque Orientale*, Paris, 1781, I, p. 363. The excerpts, by Giovanni Fontana, were published by F.D. Prager in *Physis*, 1971, pp. 341–360. About an Arabic reference, in the XIV. century, to "the Banu Musa, Philo and AlJazari" (in this sequence): Hauser on Ibn Musa, p. 15. About the XVII. century reference: see my separate note on Vaticanus 2975, hereafter.

Bassora (Basra), a rival of theirs.[70] Ancient traditions had become very vague. Scholars and translators were quite naively expected to "leave out the superfluities and fill up the gaps," assumed to exist in the old Greek books.[71]

IV. Philonian Studies from the High Middle Ages to Modern Times

During the last quarter of the IX. century, Saracen armies conquered Sicily and parts of Southern Italy, where Greek continued to be spoken: the former *Magna Graecia.* In the middle of the XI. century, this region—as well as England—was conquered by Normans. "The Arabs, who had settled in Spain [and Persia] and had been initiated there into Greek culture, brought works of this culture to Sicily and Southern Italy. There this culture was accepted by the subsequent, Norman invaders, who turned it into a new direction. Under the influence of the strong remainders of Greek civilization and language, surviving there, the Greek sources then were sought and reviewed", in addition to the Byzantine abbreviations and their Arabic translations.[72]

It seems that polyglot *scriptoria,* at the local courts, then undertook a variety of studies, in Greek, Arabic and Latin. They produced copies of Ibn Musa's work, which now are preserved in the west. They translated the work of AlKindus of Egypt, which is forgotten in the east. They must also have produced a Latin prototype of Philo's *Pneumatica,* or perhaps different Latin prototypes, reflecting different parts of this work. Soon after 1200, the Gothic architect Villard de Honnecourt in France included his copy of a drawing of Philo's Chapter 56 in a sheet of his notebook, next to an ornamental drawing

---

[70] About their exact names: Wiedemann, ed. of Al Kindus (of Egypt), pp. 230f.; for the title of the book, *De automatis et thaumantiis:* E.F. Montucla, *Histoire des mathématiques,* Paris, 1799 (an VII)—1802, pp. 406f. M. Casiri, *Biblioteca Arabico-Hispana Escurialensis,* Madrid, 1760, I, pp. 353–356, called it *De automatis et thaumaturgicis.* The two authors AlKindus are distinguished by Wiedemann and by d'Herbelot[69]. They are confused by Casiri and his Arabic source; by Montucla; by Carra de Vaux, *Penseurs,* IV, pp. 4–6; by Sarton, *Introduction,* I, pp. 559f.; and also by the present editor, in *Physis*[69], p. 346.

[71] Heath[40], I, p. 75.

[72] J.-L. Heiberg, "Neue Studien zu Archimedes," in *Abhandlungen zur Geschichte der Mathematik,* V, p. 82 (*Supplement* to *Zeitschrift für Mathematik und Physik,* XXXIV, 1890). Also see J.-L. Heiberg, "Les premiers manuscrits grecs de la bibliothèque papale," in *Bulletin de l'Académie R. Danoise,* 1891, pp. 305–318; C.H. Haskins, *Studies in the History of Medieval Science,* New York, 1960, pp. 141–222.

of the device now called Hero's Fountain.[73] Unknown writers produced other Gothic fragments, which they evidently derived from Philo's Chapters 55 and 57; they will be translated hereafter. Soon after 1200, similar studies also produced the earliest Latin copies of Philo's Chapters 1 to 23, which now exist, and another Arabic variant: AlJazari's ornamental, pneumatic drawings in his work, *Mechanical Devices*.[74] The illustrations of AlJazari (also known as AlDjazari or AlGazari, and sometimes cited without the prefix Al-) show further developments, beyond those of Ibn Musa, of constructive techniques.

Neither these eastern works nor the work of Villard show any renewed interest in the ancient pneumatic theories; but renewed or continued interest in theoretical physics appeared in Scholastic writings. Thomas Aquinas and his school (in the time of the Arabic-Norman scriptoria, of Villard, and of AlJazari), produced the first reliable, Latin translations of Aristotle, in contrast to the "filled up and corrected" variants which the Graeco-Arabic effort had produced. This fact is well settled in the history of the Aristotelian tradition.

It is less clear whether the Scholastics also influenced the Latin translations of the pneumatic books, and thereby influenced the modern renewal of Hellenistic experimental science.[75] As is well known in the history of medieval science, the very meaning of the word "experiment" underwent strange deformations. The Scholastics generally accepted and proliferated the Aristotelian dialectics about the vacuum, but some of them were aware of the other, ancient theories. Albert the Great (XIII-2 A.D.) recognized that Greek authors had explained the water-lifting phenomena, of clepsydra style, as caused "by the force of air, which supports the water." Albert, as a convinced Aristotelian, rejected this correct explanation. His contemporary, Roger Bacon,

---

[73] Villard de Honnecourt, Sketchbook, ed. T. Bowie, Bloomington, 1959, Chapter XVII (pp. 66f.). Also see Kyeser[3], fo. 95b, 112b 3, 114 v.

[74] Montucla[70], I, p. 404; Carra de Vaux, *Penseurs*, II, p. 173 and Wiedemann-Hauser on AlJazari, Drinking Vessels, pp. 56–93, 268.

[75] This general question has been discussed, from various viewpoints, by Thorndike, II, pp. 517–691 (on Albertus Magnus, Aquinas and Bacon); Grant[64], pp. 656–668; and C.B. Schmitt, "Experimental Evidence for and against a Void: the 16th Century Argument," in *Isis*, LVIII, 1967, pp. 352–366, the latter being rather confused for example when stating, p. 359, that Aristotle used the clepsydra sieve, against Anaxagoras, "to show that air has weight"). It is often asserted that the new origins of experimental science are attributable to Roger Bacon, or to his teacher (see A.C. Crombie, *Robert Grosseteste*, Oxford, 1953) or to his pupil Nicole Oresme. The questions are in need of review in the light of detailed studies of Philonian, Vitruvian and Heronian traditions.

in his *Questions about Aristotle's Four Books of Physics*, pointed out that the force of air, supporting the water (along with the smallness of the sieve's perforations) is the "effective cause" of the phenomenon; he described the impossibility of a void as "[only] the formal cause."[76]

The later Scholastics disregarded the pneumatic explanation of this phenomenon, which Albertus had reported and Roger Bacon had accepted. The views of Bacon were suppressed.

At the same time, by contrast, authors of technical, non-Scholastic notebooks gave full consideration to the ancient, pneumatic demonstrations. One such author was Konrad Kyeser (c. 1400). He was much impressed by bath furnaces of "The Great Philo" and his "Philoneus"-aeolipiles. Another such author was Giovanni Fontana, who excerpted the "book of AlKindus," the Egyptian, on fountains, pumps and pneumatic devices, including organs (c. 1418). These authors also tried to develop the tradition of military devices, found in the books of Philo and his successors.

The old, Philonian combination of leverage-type mechanics, fluid physics and military engineering, all exhibited by tangible machines, also dominated the thinking of Mariano Taccola, c. 1425–1450. His works began with recollections of harbor defenses, traceable to Philo's harbor book; they continued with medieval variants of the war engine book, and with mechanical leverage and hydraulic pneumatics, the latter including clepsydra sieves, siphons, a pipette and a water-lifting candle; they alo mentioned new experiments in the field of harbor building and hydraulic construction more generally.[77] In addition, a correct, Italian translation of the Latin Philo occurs at the beginning of one of the many copybooks produced in the middle and later 1400's and almost entirely based on Taccola's teachings. This version of Philo's pneumatics may be the first translation of an ancient technical book into a modern vernacular.

The second half of the 1400's brought the elaborating studies of Taccola's successor, Francesco di Giorgio Martini. It also brought the achievements of Leonardo da Vinci, who incidentally, acquired his own copy of the supposed Philonian book On Waters. Leonardo applied his fabulous powers of visualization to the problems of physics that he learned or discovered. Contemporary pneumatics—that is, Philonian pneumatics—was thereby absorbed into the technical and scientific Renaissance. Later engineers concerned themselves with

[76] Duhem[14], VIII, pp. 17f., 131f., 134–140.

[77] Taccola, *De ingeneis*, Books I, II and S, in his "Notebook" CLM 197, Bayerische Staatsbibliothek, Munich. Fo. 1v to 20r *passim;* fo. 68r, 69r, 73v, 75r, 134r: partly reproduced by F.D. Prager and G. Scaglia, *Mariano Taccola and his Book, De ingeneis*, Cambridge, Massachusetts, 1972, pp. 39–47; discussed pp. 34–149.

it; they studied mainly the problems of water lifted by "fire," which still were unsolved, Philo's Chapters 7 and 8. Through their study, the formerly unused potential of the Philonian thermoscope then began to be realized for scientific observation. This process also brought the first widespred interest in the work, left by Hero, on pneumatics and "the void", and the independent continuation of this work by Della Porta.[78]

Galileo then produced his *Discourse on Two New Sciences* (1638). Unnamed artisans had reported that the siphons and suction pumps, then used in attempts to lift water over mountains, could lift it only to 32 feet. Beeckman of Dordrecht and Baliani of Genoa explained this by rejecting, with Hero, the Aristotelian-Scholastic *horror vacui*; by reasserting atmospheric pressure; and by adding their beginning recognition of the natural limitation of this pressure. Galileo agreed only with that part of the new opinion which rejected the *horror vacui*. He believed that water, lifted in a suction pipe, acts as a "rope," which tears asunder when it grows to a height beyond 32 feet, but which has cohesion until then, by the mingling of atoms in distributed voids.[79] Philo had given the same, unfortunately erroneous explanation in his Chapter 3. He had been silent only about the limitation of the cohesive force.

Similarly, Empedocles, Anaxagoras and mainly Aristotle had been silent about the natural limitation of atmospheric pressure, one of the fundamental facts of modern pneumatics. However, pertinent studies had been made by a mathematician and observer, in the middle ages: Aben Moat had tried to measure the height of the atmosphere, by computing the height of the highest observable clouds, about the year 1000. His work was Latinized in the 1200's as *De crepusculis et nubium ascensionibus*, "*On Twilight and the Height of Clouds.*" It was printed repeatedly in the 1500's, and became a topic of discussion in the

---

[78] About Taccola's successors: Mariano Taccola, *De machinis*, ed. G. Scaglia, Wiesbaden, I, 1971, pp. 25–55; Prager-Scaglia[77], pp. 191–204. About Leonardo's versions of Philo's Chapter 8, see the notes to that chapter; about his book-list, citing "libro di Filone De aque:" L. Reti, *Burlington Magazine*, 1968, p. 81. About Fludd and Van Helmont: Schmidt, Thermoskop, *passim*. About the translations of Hero's "Spiritalia": Schmidt, *Supplementum*, pp. 81–100, 127–133. For G.B. Della Porta, see his work *I tre libri de' spiritali*, Naples 1606, pp. 5–61.

[79] About Beeckman: Gerland[14], p. 418; same with F. Traumüller, *Geschichte der physikalischen Experimentierkunst*, Leipzig, 1899, pp. 113–122. About Baliani: G. Govi, "il primo scopritore della pressione atmosferica," in *Atti, R. Accademia delle scienze di Torino*, II, 1867, pp. 562–568 and Caverni[14], I, pp. 436–441; VI, pp. 107–132, 211f., 219f. For Galileo, see his *Discorsi*, 1638: ed. nazionale, VIII, pp. 64–67. About pneumatic technology of this time, also see W.E.K. Middleton, *The History of the Barometer*, Baltimore, 1964, pp. 3–18.

Renaissance. The discussion in turn became known to Torricelli of Florence and Guericke of Magdeburg; it may have been noted also by Beeckman and Baliani.[80] This development, together with the rising influence of Philo's and Hero's pneumatics and the equally rising rejection of the Aristotelian-Scholastic dialectics, may explain the remarkable phenomenon of the early 1600's: Suddenly, after centuries of mere talk about *horror vacui*, came the almost simultaneous recognition of atmospheric pressure and of its limitation, in Dordrecht, Genoa, Florence and Magdeburg. There also followed the first successful measurement of this force, and the simultaneous discovery of the Torricellian Vacuum.

For a time, these achievements were rejected with some success by Athanasius Kircher, who rightly postulated that rarified vapors exist in the space that Torricelli claimed to be void. Erroneously, Kircher believed these vapors could conduct audible sound. With Berti he tried to prove this. He also tried to buttress his authority in pneumatics by developing "Kircher's Fountain," a variant of Philo's toy-clepsydras, which in turn was imitated by others. Kircher also wrote a historical book, wherein he traced the basic, mechanical devices to "a most ancient Egyptian builder (Cheops), . . . followed in various respects by Hero of Alexandria, Philo of Byzantium, and Pappus of Alexandria," [in this sequence]. The book described and illustrated supposedly Philonian machines for hauling heavy stone blocks, levers for lifting the blocks, and mechanized wedge devices for working on them; it then outlined the "wondrous uses of fluid physics." It seems possible, and somewhat probable, that Kircher knew of the "Book of Alkindus," and also of the contemporary copying of the Latin Philo, in the Collegio Romano.[81] In its intention,

---

[80] On meteorology from Aristotle to c. A.D. 1000: W. Cappelle in Pauly-Wissowa, *Supplement* VI, pp. 315–358. The *De crepusculis* (by Aben Moat, as shown by A.I. Sabra, *Isis*, LVIII, 1968, pp. 77–87) was Latinized in or about the XII. century and was printed 1542 and 1572: Sarton, *Introduction*, I, pp. 721f. Subsequent computations of the height of the atmosphere were cited by Torricelli (letters to Ricci, 1644, see G. Loria, ed. of E. Torricelli, *Opere*, Faenza, 1919–1944, I, pp. 472–476, quoted by Rouse-Ince[5], p. 63) and by Guericke (*Experimenta nova*, ed. Schimank and Krafft, Düsseldorf, 1968, Book 5 Chapter 7).

[81] On the Kircher-Berti experiment: A. Kircher, *Musurgia universalis*, Rome, 1650, pp. 10–12; G. Schott, *Mechanica hydraulico-pneumatica*, Würzburg, 1657, pp. 306–309; Gerland-Traumüller[79], pp. 10–12. The view of De Waard[14], pp. 101–110, 116f. is pertinent but questionable. On "Kircher's Fountain:" A. Kircher, *Magnes*, Cologne, 1643, pp. 523–525; Schott, *op. cit.*, pp. 12, 177f.; S.A. Bedini in *Physis*, X, 1968, pp. 25–52. For views on Philo and other "most ancient Egyptian builders:"

Kircher's entire performance was an attempt to defend the Aristotelian system.

As is generally known, that system then collapsed. We must also note that Philo then fell into oblivion. One reason of the latter fact may lie in the greater plausibility of the vacuist demonstrations developed by Hero (rediscovered shortly before) and by Torricelli. The "scientific revolution", brought about by the Torricellian achievements, was over-estimated to the extent that the ensuing Enlightenment dismissed all former science as a succession of "errors and chimeras."[82]

However, Philo's work was not irretrievably lost. Thévenot and DeLaHire produced a beautiful *editio princeps* of his and his pupils' works, known at the time. The edition was published in 1693 as *Veterum mathematicorum opera*, in Greek and Latin. It was illustrated by Sédileau in drawings copied from medieval manuscripts, including the challenging, pneumatic automaton, copied in the frontispiece of this work. Through the publication of Philo's books on war engines, it became widely known that he had also written a book on pneumatics although the surviving copies of this book gathered dust in the archives.

In the 1700's, Johann Albert Fabricius made it fairly clear how the physico-technical views of classic Greece had been developed by the Alexandrians, and had then been preserved in part by the Byzantines. Fabricius' work became deservedly famous among scholars of literature, but was not noted or paralleled in the great, contemporary history of the exact sciences, written by Montucla, or in the pertinent articles of the French *Encyclopedia*.[83] Those works showed only some slight, pertinent interest in Hero. It may be due to this fact that the interest in Philo and Hero still is disproportionate in the modern encyclopedias, such as the Britannica and the Italiana.

Studies of further linguistic scholars led to the gradual rediscovery of Philonian manuscripts. In the early 1800's the Arabic "book of excerpts" found some interest. Then, in the middle of the century, a most significant article about Philo was published. Its author, Friedrich Haase, had conducted extensive

---

A. Kircher, *Oedipus aegyptiacus*, Rome, 1664, pp. 312–325, with drawings attributed to "Philo's book *On the Five Static Powers*," considered also by De Rochas, 1912, p. 61.

[82] Voltaire, *Le siècle de Louis XIV* (1751), in his *Oeuvres historiques*, ed. Paris, 1957, pp. 997f. Also see Kant: science had only been "groping around" until Galileo and Torricelli, who made "a light dawn on all explorers of nature." *Kritik der reinen Vernunft*, preface to the second edition (1787): *Gesammelte Schriften*, III, Berlin, 1911, p. 10.

[83] J. A. Fabricius, *Bibliotheca graeca*, ed. Hamburg, 1790–1830, IV. pp. 231f. Montucla[70], I, p. 268; D. Diderot and J. l. R. D'Alembert, *L'Encyclopédie*, 1751–1772, s. v. Pneumatique.

studies of manuscripts, which enabled him to give a new and superior analysis of Philo's works. He tried to publish the manuscripts but unfortunately found this impossible. His article on Philo has remained unnoted by newer generations of scholars, although his collection of materials was inherited by them.[84]

In 1870, Valentin Rose revolutionized the Philonian studies by announcing his discovery of the Latin *Pneumatica* of Philo. In his brief and brilliant article he published the text of a London-Sloane manuscript, which he collated with manuscripts in Munich and Paris. Intuitively he postulated that the Latin text had been translated from a lost Arabic prototype, as it begins with an Islamic-sounding invocation of God. In two plates, appended to the text of Philo, Rose published drawings of Philo's devices, which a professional draftsman had produced for him as freehand copies from the London and Munich manuscripts.[85]

A significant contribution was then made by Diels (1893), who discovered a parallelism between Strato, Hero and others, in their remarks about the "void". Deals recognized that the experimental practice of the Alexandrians—also in medicine—had beginnings in the Aristotelian school. As a "pure" classicist, he dismissed Philo and Hero, the authors of later and more empirical monographs, as "mere" artisans.[86]

Carra de Vaux then discovered, among other things, the Arabic *Pneumatica* of Philo, and published it in 1903. His publication greatly enlarged the basis of Philonian studies.

The works of Rose and Carra de Vaux led to such studies, and to further translations, including those of De Rochas, Gerland-Traumüller and Schmidt before the turn of the century, and thereafter those of Beck and Wiedemann-Hauser.[87]

The publications of the texts, mainly by Schmidt and Carra de Vaux, have hardly been questioned, but there has been little satisfaction with the drawings

---

[84] Th. H. Martin, "Recherches sur . . . Héron," in *Mémoires*, *Académie des Inscriptions*, ser. 1, IV, 1854. Haase, pp. 428–435. For the use of Haase's unpublished material by the next generation: Diels, Straton, p. 120; Schmidt, Introduction to Hero, I, p. XXVI. For a bio-bibliography of Haase: Sandys[60], pp. 137f.

[85] Rose, II, pp. 281–314, based on manuscript **S** collated with **M, N** and **P.** The editor, V. Rose, should be remembered also for his edition of another illustrated text: *Sorani Ephesii vetus translatio latina*, Leipzig, 1882.

[86] Diels, Straton, p. 110n: "reine Banausen." Similarly already H. Cardanus, *De subtilitate*, Nürnberg, 1550, p. 5.

[87] De Rochas, 1912, pp. 195–208; Gerland-Traumüller[79], pp. 33–44; Schmidt, ed. of Hero, I, pp. 458–489; Beck, pp. 64–77; Wiedemann-Hauser on Irun's Extracts, pp. 147–154; Diels-Schramm, eds. of Philo's *Belopoiica*, 1918, pp. 3f.; 1919, pp. 3–14, on Philo's style.

illustrating those texts. The Carra de Vaux edition uses modernized variants of the authentic drawings, while the Schmidt edition uses sketches by Haase, derived from unspecified manuscripts. This kind of illustration constitutes a step backwards from the standard established by Rose, particularly as photocopying techniques have been developed, which call for an edition with accurate reproductions of the authentic drawings.

None of the text editions published thus far pays attention to the results of archaeological excavations. A few men, generally interested in the history of pneumatics, have tried to clarify this history by consideration of pertinent artifacts.[88] Mainly the systematic report by Deonna (1908), showing the principal findings of the decades about 1900 and discussing them in the light of the statements of Philo and Hero, is very useful for the study of the record.

Only the purely textual studies have been considered in the more general works,[89] and some of their contents have been misinterpreted. For example: Carra de Vaux, in 1903, considered it "probable" that certain pumps and the like, shown in the Oxford "book of extracts", should be attributed to Philo. This cautious opinion has been converted into a definite, factual statement, attributing these pumps to Philo.[90] In a work of 1921, Carra de Vaux stated his position even more cautiously, and no longer called the attribution even "probable."[91] This latter view has remained rather unnoted.

Some of the most interesting studies of Philo and Hero were published by Miss Hammer-Jensen. As to Philo, her studies successfully challenged the opinion of Diels.[92] Her views in turn were attacked, in regrettably intemperate form, by Hoppe.[93] At least as to Philo, her conclusions remain quite unshaken.

---

88 Ewbank[6], *passim*; Deonna, *passim*; M. Schmidt[11], pp. 11–30.

89 For example: E. Mach, *The Science of Mechanics: A Critical and Historical Account*, ed. La Salle, 1960, pp. 130–133. B. Gille in Singer, II, pp. 633, 636. A.R. Hall[46], pp. 708–713. A. Kenny, "Vacuum Theory and Technique in Greek Science," *Transactions, Newcomen Society*, XXXVII, 1967, pp. 47–53; M. R. Cohen and I.E. Drabkin, *A Source Book in Greek Science*, Harvard, 1958, pp. 245, 255f., 334f., 346f.

90 So for example by Drachmann, p. 45; also by the present editor in his note on Fontana[69], p. 352; also Cohen-Drabkin[89], p. 255.

91 Carra de Vaux, *Penseurs*, II, p. 171.

92 Mainly by considering Philo's theories and demonstrations as anticipating those of Hero: pp. 413–427 and 480–498; also see "Heronische Frage," pp. 36–40. However, the influence of Diels and of eclectic classicism persists. So for example when Sarton writes, I, x: "Greek culture is pleasant to contemplate, because of . . . the absence of gadgets."

93 E. Hoppe, in *Hermes*, LXII, 1927, pp. 67–98.

The book of Ibn Musa has been published by Hauser (1922), and Philo's work has been reviewed by Drachmann (1948) and recently by Krafft (1972). The studies of Hauser and Drachmann suffered from the lack of access to authentic manuscript drawings, as the authors recognize.[94] They also overlooked the archaeological record. However, Drachmann and Krafft have analyzed the texts in masterly ways. Their suggestions will be cited at many points and discussed at a few, in this new edition.

---

[94] Hauser on Ibn Musa, p. 19; Drachmann, pp. VIIf., 41f.

Documentary Details of the Manuscripts

The following manuscripts are reported by historians, bibliographers, and archivists as containing Philo's *Pneumatica* or parts of this work.[100] Except as shown, all are in Latin.

Cambridge, Pembroke College, Ms. 169, first two folios (7½ columns), not paginated. Cited as Cambridge or **C.**

Florence, Biblioteca Laurenziana, Ashburnham Ms. 210, fo. 141v–144v. Cited as Ashburnham or **H.**

Florence, Biblioteca Nazionale, Fondo Conventi Soppressi, Ms. C.S.J., IX. 26, fo. 67r–70v. Cited as Fondo Conventi or **F.**

Geneva, Bibliothèque Publique et Universitaire, Ms. 80, fo. 28r–32r, 61/62. Cited as Geneva or **G.**

Istanbul, Ayasofya Museum, A.S. 3713, fo. 1a–86b, cited as Ayasofya or **A.** This manuscript, its later copy A.S. 2755, fo. 91b–146a, and a still more recent copy in Hyderabad,[101] are in *Arabic.*

Kraków, Biblioteka Jagellońska, Ms. 568, fo. 104r–106v. Cited as Krakow or **K.**

London, British Museum, Additional Ms. 34113, fo. 1r–8r, in *Italian,* cited as Italian Philo or **I**;
Egerton Ms. 2622, fo. 85r–88v, cited as Egerton or **E**;
Sloane Ms. 2030, fo. 110r–114r, cited as Sloane or **S.**

Munich, Bayerische Staatsbibliothek:
CLM 444, fo. 213r–214r, cited as Munich or **M;**
CLM 534, fo. 26v–29r, cited as **N.**

---

[100] Mainly L. Thorndike and P. Kibre, *A Catalogue of Incipits of Medieval Scientific Writings in Latin*, Cambridge, Massachusetts, 1963; P.O. Kristeller, *Iter Italicum*, London, 1963–67, both s.v. Philo; and G.D. Macrae, *Catalogi Codicum Manuscriptorum Bibliothecae Bodleianae, Pars IX*, Oxford, 1883 (col. 36 on Digby 40).

[101] M. Krause, in *Quellen und Studien zur Geschichte der Mathematik, Astronomie und Physik*, B. Studien, 1936, III, pp. 442f.

Oxford, Bodleian Library, Eastern Division: Ms. 954, fo. 1a–48b, in *Arabic*, cited as "Book of Extracts" or **B**;

Oxford, Bodleian Library, Western Division:

Ashmole Ms. 1471, fo. 167v–168v, cited as Ashmole or **R**;
Digby Ms. 40, fo. 9r–12v, cited as Digby or **D**;
Lyell Ms. 36, fo. 12r, v, cited as Lyell or **L.**

Paris, Bibliothèque Nationale:

Cod. Lat. 7295, fo. 141v–142v, cited as Paris or **P.**

Prag, University Library:

Cod. 1832, fo. 130r–133r, cited as Prag.[102]

Rome, Collegio Romano:

Cod. H.C. 93 (Vittorio Emmanuele 2548), fo. 92r–96v, cited as **U.**

Vatican City, Biblioteca Apostolica:

Cod. Vaticanus Latinus 2975, fo. 210r–214v, cited as **V**;
Cod. Palatinus Latinus 1332, fo. 92v–95r, 119v, cited as **W**;
Cod. Palatinus Latinus 1369, fo. 140r–141v, cited as **X**;
Cod. Palatinus Latinus 1375, fo. 180r–181r, cited as **Y**; and
Cod. Palatinus Latinus 1377, fo. 5r–9r, cited as **Z.**

Three of these documents are reproduced in this edition: a typical, Latin manuscript, CLM 534 (**N**); the recently discovered Italian Philo, Add. Ms. 34113 (**I**); and an English version of the principal Arabic manuscript, A.S. 3713 (**A**). This Arabic manuscript is, in many respects, the most comprehensive copy of the work. It is a book-filling treatise of 172 pages, in small octavo format, with a drawing on every second or third page of the text. Other and smaller documents are described in the following, documentary notes.

A NOTE ABOUT TERMS, USED IN THIS EDITION: An attempt is made to determine the nature of Philo's "*original*" work, in contrast to the later "*archetypes*" and "*prototypes*", and to the existing documents. This edition designates all of these existing documents as either "*copies*", or "*excerpts*" (fragmentary copies), or "*variants*" (modified copies, also called "*versions*" when the modifications are slight). Of course it recognizes that copies, and mainly diplomatically accurate copies, are most significant for our attempt to reconstruct the source documents (prototypes, archetypes and originals), but that variants,

---

[102] A. Birkenmajer, in *Beiträge zur Geschichte der Philosophie des Mittelalters*, XX, 5, pp. 30f.: the Prag manuscript is copied from **K.**

and mainly those which indicate a technically correct criticism, are most important for our attempt to determine the history of scientific progress.

We may generally assume that Philonian originals and archetypes were written in Greek, and may designate all "*translations*" as prototypes of derived, existing documents (copies, excerpts and variants) in other languages; actually, in Arabic and Latin. We will assume that the modifications, contained in variants, stem either from other documents, or from artifacts, or from improvements or deteriorations in the copying process.

In the study of the record we encounter actual improvements, as well as various types of intended, erroneously assumed improvements, of technical or scientific nature, while we see the copies and plain translations substantially preserve the earlier message. It is the main object of these documentary notes to review the Philo manuscripts with the question in mind to what extent they appear to be copies or variants of the original.

I. Philo's Technology and Science according to the Latin and Arabic Traditions

At some points it is rather obvious that an Arabic interpolation is present. For example we can safely assume this when a device, as in Chapter 63, is developed for a temple or a "*mosque*." It is less easy to say whether the latter, non-ancient term is a substantial addition to the source document or only a semantic amplification of its translation.

The Arabic text often uses measures which bear non-ancient names, such as *qist* or *ritl*, as in Chapters 11, 14, 15, 18. These could be pure translations, derived from the Greek source document; the Greek language indeed had related terms, such as *xestos* and *litron*. However, these terms could also be independent additions. Thus it remains an open question whether the imaginative pseudo-accuracy of these terms is Arabic in origin, as it often is in other cases,[103] or is taken over from the Greek style of Philo's original, or of the archetype editor.

Some proposals of the technology or science, shown in the existing Philo copies or versions, are not reflected in other ancient books, presently known. We cannot conclude as a matter of course that such proposals are later interpolations. We may have to revise our view of early, ancient developments. What further evidence is needed, or admissible? For example, when we find that a Philonian device was not used in Roman practice, does it follow that

---

103 W. Hinz, *Islamische Maße und Gewichte*, Leiden, 1955 (*Handbuch der Orientalistik, Abt. I, Ergänzungsband* I, 1), Vorwort.

it had no influence on ancient theory? or that its description was no longer copied in later antiquity or in the early middle ages?[104] Was Philo ahead of his time? Was his influence an indirect one? Occasion will arise, in notes to the translated text, to indicate different answers to these questions.

It is usually assumed that the ancient, Greek originals of Philo's books on mechanics and pneumatics are lost. In a strict sense this is true. However, we have Greek fragments. We shall cite them, and consider whether their style, or their content, has historic significance.

It has also been assumed that an early Arabic prototype, not the direct use of a Greek document, gave rise to the Latin translation that was rediscovered by Rose. Such was the case in a number of other works of ancient literature. Also, a special proof of this assumption has been found in the use of an Islamic, introductory formula, in the Latin translation of Philo's Chapter 1. However, *problems of the assumed, Arabic-Latin descent* face us at many points, when we consider the scientific-technical contents of the book, rather than its religious *initia*. At this point we will consider some of the problematic elements.

First, we should suspect a "degeneration of realism" when drawings and their descriptions are copied from books, and when the copyist has no recourse to observation of real things, or at least of authentic copies. There was very pronounced "degeneration" of this type in the medieval copying of ancient natural science.[105] None of the usual indicia of "degeneration of realism" has been noted between the Arabic and the Latin copies of Philo. On the contrary, at least one of the Latin manuscripts (the one in Cambridge) shows greater realistic veracity than the Arabic equivalent, in some of its copywork.

Second, when studying the Arabic and Latin versions of the first score of chapters—the set which the two versions generally share—we find large *lacunae* in both versions, but the Arabic *lacuna* is not at all reflected in the Latin translation. Rather, the situation suggests that the Arabic and Latin translations were founded on Greek archetype materials of deteriorating kind; of different, physical conditions at the times of the different translations; and not necessarily of better condition at the time of the Arabic translation. The evidence relates to Chapters 9 to 12 (Arabic) and 12 to 16 (Latin); it will be considered in detail in connection with those chapters.

Third, the general, technical character of the Latin translation differs from that of the Arabic translation, in the areas unaffected by *lacunae*. The character of the Latin translation is more plausible, as an exercise in "pneumatics". The defect of

---

[104] So for example Drachmann, *Technology*, p. 190; Krafft, p. 344. The other conclusion, mentioned hereafter, is that of Jähns[25], p. 42; also see Haase, p. 431.

[105] Mason[4], p. 109.

the Arabic variant is mainly due to the Arabic chapters which follow the first score. Those later chapters contain propositions from different sciences, mixed with a few further pneumatic demonstrations, while the first score is purely pneumatic.

Predecessors of Philo had distinguished the pneumatic forces from the hydraulic and mechanical ones. As a result, the importance of a distinction between such matters was understood, at least by scientists.[106]

Philo can pass as a man of scientific tendencies. His military books, surviving in Greek, indicate single-minded concentration on the subjects defined by their titles.[107] The first score of chapters in the present work is of similar nature. By contrast there is frequent, unexplained shifting of interest, in the last two scores of chapters of the Arabic translation. The impression arises that those chapters—partly reflected also in Latin-Gothic fragments attributed to Philo—belong to later compilations of Philonian works and works by others. Evidence of this matter will be noted hereafter.

Current views about Philo's book disregard most of these elements, and rely entirely on the Islamic style of an introductory note to the Latin translation. Such influences of style, however, can be explained in different ways, as will be shown in a special note to this documentary problem.

Aside from the *lacunae*, mentioned above the manuscripts present pneumatic, hydraulic and mechanical topics in form of small groups of chapters, each group containing four or eight chapters. At least such is the case in some parts of the manuscripts.[108] We might raise the question of Philonian authenticity of this method of presentation. We find that no such method is used in his relatively more authentic, Greek books on military engineering, and that he, when dedicating his pneumatics book to this friend, merely speaks of "different devices." It seems most likely, for this reason, that the four-chapter grouping was introduced by a later editor.

II. Historic and Paleographic Dates

The general historic setting of the Philonian record is summarized here in Chart 1, and the sequences of more specific documents are listed in Chart 2. Their reported dates are shown thereafter, with so much detail as the existing, paleographic estimates allow.

The Latin documents are written in calligraphies which range from the early beginnings of the "Gothic" script to "Humanist" versions. In most cases

[106] Aristotle, *Meteorologica*, I, 13, p. 349b35; IV, 2, p. 379b14–18; IV, 3, p. 380a 17–30; *Historiae animalium*, p. 586a17; *Partes animalium*, p. 625b19; *Problemata naturalia*, p. 916b4, 6, 7 and 917a20, 22.

[107] Haase, p. 430.

[108] Drachmann, p. 69.

a "broken letter style" is used for the text in general, with occasional insertions in either "cursive" writing—mainly for invocations and dedications—, or in "book hand"—mainly for subtitles.—The eastern manuscripts use writing of a "secular" type.

A word may be added here about the earliest, surviving, Greek copies of Philo's military books. These were written at somewhat earlier dates than the translations of the *Pneumatica*; some of the Greek calligraphy is of a time about A.D. 1000.[109] The Latin copies of this pneumatic work, preserved now, began to be produced shortly after 1200, and the Arabic copies, sometime after 1300. These latter dates are not necessarily the dates, or close to the dates, when the Latin and Arabic prototype-translations were made, but they may be considered suggestive for the relative dates of these prototypes.

For general orientation about this matter, we may also note the range of dates generally attributed to the existing copies and the prototypes of other scientific works, which originated in Greek antiquity. The translations of the Aristotelian *Organon* began in late antiquity, with Graeco-Latin interpretations; they continued with Arabic-Latin versions (sometimes Arabic-Hebrew-Latin versions) during early Scholasticism; and they then led to the more direct and more accurate Graeco-Latin translations of mature, western Scholasticism. Regarding the *Elements* of Euclid, the sequence was similar.[110] Substantially the same thing is true of basic works by Archimedes and Ptolemy; as to these works, most significant contributions were made by the polyglot Sicilian scriptoria.[111] Philo's *Belopoiica*, with possible admixtures from his *Mochlica*, appears in Arabic excerpts and versions, in the time of the Banu Musa. For Hero's and pseudo-Hero's *Pneumatica* books, neither an Arabic translation nor even an Arabic version became generally known, but Ahmed ibn Musa must have relied on a Persian copy. The Greek text of a Heronian book "on the vacuum" was known in Sicily in the XII. century, and was Latinized then or at least beginning in the XV. century.[112]

---

[109] London, Burn. Ms. 69; Paris, cod. grec. 2445; Vatican City, Vaticanus graec. 1164.

[110] About Aristotle: Sarton, *Introduction*, II, pp. 179f., 561–568. About Euclid: ed. Heath[40], I, pp. 75–79, 91–96; Sarton on Euclid and Ptolemy, pp. 3–36.

[111] On Archimedes: Heiberg, Neue Studien[72], pp. 80–88; M. Clagett, *Archimedes in the Middle Ages*, Madison, 1964, I, pp. 3f. On Ptolemy: Thorndike, II, p. 659; Sarton on Euclid and Ptolemy, pp. 37–73.

[112] Schmidt, in Hero, *Supplementum*, pp. 49–53. The Latin translation is by G.B. Burana. Also see Grant[64], pp. 656–657. About the Arabic extracts from *Belopoiica* and *Mochlica*: Hauser on Ibn Musa, p. 17.

Chart 1. Basic Names, Documents and Dates of the History of Philo's *Pneumatica*

	Greek Tradition	Latin Tradition	Eastern Tradition	Italian Tradition
–500				
	Empedocles			
	Anaxagoras . . .			
–400	Democritus . . .			
	Aristotle . . .			
–300	Strato . . .			
	Ctesibius			
–200	*Philo*			
–100				
0		Vitruvius		
100				
	Hero			
200				
300				
	Pappus			
400				
500				
	Proclus		Approximate	
600			Time of Persian	
700	Hero of	Celtic and	Translation	
	Byzantium	Carolingian		
800		Renewals	Hellenistic-	
			Arabic Renewal	
900				
	First Greek			
1000	*Belopoiica*			
	Copies extant			
1100				
1200	Tzetzes	First Latin Philo		
		Copies extant		
1300				
			First Arabic	
1400			Philo Copies	
			extant	Italian Philo
1500				Copy
				Della Porta . . .
1600		Kircher		
				Galileo . . .
				Torricelli . . .

Arabic work was dominant, in the west, in all sciences where Greek beginnings were scarce or lacking: mainly in algebra, trigonometry, and some parts of optics. In pneumatics, by contrast, the Greek beginnings were plentiful. In early Byzantine times they may have been known from Ireland to Persia. Therefore, *the current theory is in need of proof when it asserts a priority of an Arabic translation of Philo's Pneumatica text and illustrations.* It seems possible and somewhat probable that the existing Latin Philo descended basically from work which comprised the Latin Irish contributions of pre-Arabic times. This remains quite possible also if Arabic elements, such as an invocation, later filtered into the Latin Philo. Latin Philo copies became possible in Vitruvian times. Irish copies as well as Persian translations became possible a very few centuries later. Arabic ones became possible, but are not known, from the time of the Banu Musa.

Chart 2. Dates of Different Philonian Traditions

Approximate Years A.D.	Celtic-Gothic Tradition	Latin Tradition L-II	Latin-Italian Tradition L-II	Persian-Arabic Tradition
700– 800	*Carolingian Prototype*			
800–1000				Ibn Musa, AlKindus
1000–1200		*Sicilian Prototype*		
1200–1300	Villard	**D; C**		AlJazari
1300–1450	Kyeser	**N, S, W, X, Z**	**P; I**	**A; B**
1450–1600		**F, G, H, R; K**	**E, L, U, V, Y**	

The entries in cursive type, in Chart 2, are hypothetical. The bold-faced entries are based on paleographic estimates, which are as follows:
XIII. century, early part: Digby 40 (**D**).[113]
XIII. or XIV. century: Cambridge-Pembroke 169 (**C**).[114]
XIV. century: A. S. 3713 (**A**); CLM 444, 534 (**M, N**); Sloane 2030 (**S**).[115]

---

113 Macrae[100], col. 36; A. A. Björnbo and S. Vogl, "AlKindi . . . Werke," in *Abhandlungen zur Geschichte der mathematischen Wissenschaften*, XXVI, 1912, pp. 132, 134.

114 M. R. James, *Descriptive Catalogue of Manuscripts in Library of Pembroke College*, Cambridge, 1905, p. 163; Thorndike-Kibre[100], col. 694.

115 On **A:** Krause[101], p. 442; on **M, N, S:** Rose, pp. 288f.; on **M, N** also Thorndike-Kibre[100], col. 696. The estimate of Rose appears more reliable than that of the B. M. manuscript catalogue, which speaks of the XII. or XIII. century.

XV. century, years unknown: A. S. 2755;[116] Bodleian 954 (**B**);[117] Egerton (**E**), Fondo Conventi (**F**), Ashburnham (**H**), Lyell (**L**), Paris (**P**), Ashmole (**R**, Palat. 1332 (**W**), Palat. 1377 (**Z**).[118]

XV. century, middle: Palat. 1369 (**X**); Add. 34113 (**I**).[119]

XV. century, end: Palat. 1375 (**Y**); Krakow 568 (**K**); Prag.[120]

XVI. century: Geneva (**G**);[121] Collegio (**U**);[122] Vaticanus (**V**).[123]

III. Recorded Names of the Author and of Scribes

1. The Author's Names

The Greek name, Philon, is most accurately copied by Martini. It is transcribed in Arabic as
"Filun" in **A** and **B**; in Latin as
"Philo" in **C, D, P, Y** and in a chapter of Kyeser;
"Phillo" in **N**;
"Philonius" in **F, H, R, W** and the explicits of **X** and **Z**;
"Philōníus" in **M**;
"Filonius" in the explicit of **G**;
"Filo" in **U** and **V**; and in Italian as
"Filone" in **I**.
No name is given in **E** and **L**. The *Pneumatica* is attributed to
"Aristotiles" in **S**[124] and in the incipit of **Z**, and to
"Philo Hebraeus" in **K** and Prague.[125]

---

116 Krause[101], p. 443.

117 Carra de Vaux, p. 29n.

118 On **E**: A. De la Mare, *Catalogue of the Lyell Manuscripts*, Oxford, 1971, p. 94. On **F**: Kristeller[100], p. 165. On **H**: Thorndike-Kibre[100], col. 456. On **L**: De la Mare, *loc. cit.* On **P**: Rose, p. 289. On **R**: W. H. Black, *Descriptive . . . Catalogue of Mss. of E. Ashmole*, Oxford, 1845, col. 1280. On **W**: Kristeller[100], p. 393. On **Z**: Kristeller[100], p. 393; Björnbo[113], p. 134.

119 On **I**: see the note on the Italian Philo, below. On **X**: Kristeller[100], p. 393.

120 On **Y**: Kristeller[100], p. 393; Thorndike-Kibre[100], col. 696. On **K** and Prag: Birkenmajer[102], p. 23; Thorndike-Kibre[100], col. 1441.

121 Thorndike-Kibre[100], col. 1305.

122 Björnbo[113], p. 142.

123 Björnbo[113], p. 144; Kristeller[100], p. 393.

124 Rose, *Aristoteles*, p. 155.

125 Birkenmajer[102], pp. 30f.; also see Grant[64], p. 660n.

2. Scribes and Scriptoria

A Greek copy, designated in Latin as a Liber de ingeniis, was owned by a Sicilian scriptorium at the court of a Norman or Hohenstaufen ruler in the XII. or XIII. century. This scriptorium, or a neighboring one in this polyglot area, may be assumed to have produced a Latin Philo prototype, and perhaps some of the early copies from it, such as **C** *or* **D.** *The Latin translator in Sicily no doubt was exposed to Arabic and Islamic influences, in such matters as the formal initia of the work. He worked near Arabic writers. For example the scholar and "Emir" Eugenius was designated as "a man fluent in Arabic as well as Greek and Latin." Thus it is not surprising to see an Arabic copy, such as* **A,** *influenced by western drawing elements, and to see a western translation influenced by a predominantly Islamic tradition for the invocation of God.*[126]

Individual names of early copyist translators, concerned specifically with Philo's *Pneumatica*, are rarely preserved. Only one out of more than a score of copies is expressly signed, dated and located. This is copy **Y,** now in Rome, which contains the following note in red ink, at the end of the Philo text:

> Explicit liber . . . in inclito studio Heidelbergensi, anno 1493, domenica Sancti Viti, per ingenium Joannis Virdinigi de Hassiret.

This Jean de Hassiret also signed the first folio of the codex, indicating that he compiled part of it in Krakow, about or after 1520, not for local use but "illustri pro domino Lionello marchioni Estensi" of Ferrara. De Hassiret then also remarked that he was once visited by Nicolaus Schoenberg,[127] who is now remembered as a friend of Copernicus. De Hassiret is otherwise unknown.

Another version, **K,** is preserved in Krakow, and may have originated there in the late 1400's. Other manuscripts in the same volume indicate this time and place of origin. Some of them are attributable to Henricus Griffinus (Henry of Greifswald), an otherwise unknown writer who signed various parts of volume **K.** A copy of **K** went to Prag. Other copies, related to **Y,** originated at or near Heidelberg or were kept there when the collection went to the Vatican.

It is conceivable that a separate Latin prototype was developed, in the early 1100's, by Avendar (also known as Abendavit, Johannes Hispalensis, or Johannes Paulinus), a man who translated many Arabic books into Latin. Among

---

126 Heiberg[72]. Also see Haskins[72], pp. 141, 172; also O. Hartwig, "Die Übersetzungsliteratur Unteritaliens", in *Centralblatt für Bibliothekswesen*, III, 1886, pp. 161–190.

127 ". . . Nicolaus de pulchro monte, et nominatus Schonberg Britannius . . . fuit mecum anno 1520 . . .": Sheet a/A, preceding fo. 1.

his translations is that of an Arabic, alchemistic treatise by Alcanus; that translation, along with the Latin text of Philo, was later copied in various manuscript volumes, including **L** and **M.**[128] The copies are fragmentary, and the translator's personality is elusive.

IV. The "Initia"

1. The Invocations

Most of the Philo *Pneumatica* manuscripts are prededed by the formula: "In the name of the charitable and merciful God," *In nomine dei pii et misericordis.* These words are sometimes inserted in a distinctive, cursive handwriting, as for example in the Cambridge and Digby manuscripts, **C** and **D.** There are occasional modifications, such as "*misericordissimi*" in one of the Palatine manuscripts of the Vatican, **X,** as well as additions, for example "... *cui sit honor et gloria per infinita seclorum secula,*" in **C.** In a few cases (**I, K**) the formula is replaced by some abbreviation of the *Nomen Sacrum* of Christ. Quite frequently the entire formula is omitted, as in **G, H, W;** so particularly where the book is attributed to Aristotle, as in **S** and **Z.**[129]

We may wonder if it was the supposedly Hermetic character of the text which was assumed to require an invocation of divine charity or foregiveness. The formula is generally explained in a different way. *Ever since the rediscovery of the Latin Philo, the theory has been current (a) that the invocation must have been copied from an Arabic source, because similar words occur in a Mohammedan ritual known as the "basmalah." As a further conclusion it has been assumed (b) that the entire Latin text must come from a Muslim source, as in the case of many Arabic-Latin translations of Greek books.*[130] The theory was briefly questioned by Heiberg.[131] It was generally accepted after the discovery of the Arabic Philo and of its *basmalah.*

Both parts of the theory may be questioned. (a) The invocation shows that there may be an Arabic influence on the Latin text, but not that there "must" be

---

[128] Thorndike, II, p. 794; De la Mare[118], p. 94.

[129] **P** as well as **N** has an Incipit without invocation, preceding the one with it. **G, H, S, W** and **Z** have only the former.

[130] Rose, p. 229; Schmidt, introduction to the Philo translation in Hero, I, p. 459; Carra de Vaux, p. 27; Sarton, II, p. 347; Drachmann, p. 42; Krafft, pp. 344, 367.

[131] Heiberg, "Les premiers manuscrits,"[72] pp. 308, 314 describes the Greek *De ingeniis,* mentioned in an early, Sicilian library catalog, as "les *Pneumatica* de Philon, dont nous avons une traduction latine (d'après l'arabe?) ... (Rose ...)."

such influence. Similar formulas are used in Christian liturgies, and the same formula is used in the Latin text *De aspectu lune*, which is not known to be influenced by Arabic sources.[132] The Islamic use of the formula came at a late date; previously the formula had been established in Hebrew, Greek and Latin:

el	rahum	ve- hanun	: Exodus[133]
kyrios	oiktirmōn	kai eleēmōn	: Septuagint, III-1 B.C.
deus	misericors	et clemens	: Vulgate, IV-2 A.D.
'llāhi	-l-rahmāni	-l-rahīmi	: Koran, VII-1.

Even if this point be disregarded because of the greater familiarity of the formula in Islam, it still (b) does not follow that the entire Latin text comes from an Islamic source. In various centers, including Norman Sicily, religious tolerance prevailed, and there were complex cross-influences. The sponsor of a Graeco-Latin translation may have been Islamic-influenced to the extent that he placed the Islamic formula on the front page of his Graeco-Latin book.[134]

Evidently the problem is puzzling. We will return to it, and will emphasize it, at a number of points, some of which indicate a direct Greek descent, not a Graeco-Arabic one, of technical contents of the Latin Philo.

2. The Dedications

In dedicating his books to Ariston, Philo used the then accepted greeting formula: "Philo wishes Ariston joy," *Philōn Aristōni chairein*. This is shown by his military books.[135]

Latin translators misunderstood the name Ariston. The Latin copies are dedicated to strange unknowns such as "*marzotom*," "*mi argutom*," and even "*mi antoniane*." The Arabic ones speak of "*muristom*;" they use one relatively plausible form, "ristun". It has long been recognized that all these forms are corruptions of "Ariston."[136] They may enable us to deduce relative sequences of manuscripts, as it seems probable that the more corrupted forms came after the less corrupted ones, in the history of the prototypes.

---

132 The liturgical poems are by Ambrosius (c. 350) and Gregor (c. 600). *De aspectu lune* is in Vatican City, Palatinus Latinus 303, fo. 83, XIII. century (Thorndike-Kibre, col. 697).

133 Exodus 34 : 6; also Psalm 86 : 15, etc.

134 Roger of Hereford used the formula in the Incipit of an astronomical book which may be his own work *or* a translation from the Arabic; written about 1170 and copied about 1470: Thorndike, II, pp. 181 f.

135 So also Archimedes to the astronomer Dositheus, or to King Gelon: *Opera*, ed. Heiberg, I, p. 1 and *passim*; Pauly-Wissowa *s.v.* Ariston.

136 Carra de Vaux, p. 33; same, *Penseurs*, II, p. 179.

The Arabic "*ristun*" may stem from a simple contraction of Ariston; "*Muristom*" (also pronounceable in other ways, including "*maristom*") may be derived from a Byzantine-Greek form of an archetype, which could be read "*O phile mu Ariston,*" My friend Ariston.

Such a Byzantine-Greek form may also have been translated into a Latin prototype form, "*amice mi Ariston,*" which may have led to the corrupt forms, now existing. A fairly complete system of Graeco-Arabic derivations, in parallel with Graeco-Latin derivations, is presented here, in Chart 3. It uses cursive type for the hypothetical forms of the archetypes and prototypes; it uses plain roman type for the reported original, and for the existing copies.

The chart indicates the possible existence of different Latin traditions, identified as L-I and L-II. Each of these in turn includes subsidiary groups of manuscripts, which seem dependent on prototype dedications such as "*miarzatō*"-"*marcatō*"—"*miargatō*". The chart also indicates that up to four instances of corrupt copywork may lie between the first, Latin prototype and the "early," existing copy, Digby 40 (**D**), which paleographically dated in the early 1200's. Accordingly, the Latin prototype may have been produced well before 1200; it seems unlikely that the successive instances of copywork followed one another within less than a few decades, as an average.

With reference to the current theory, about the origin of the Latin prototype, we may also note what the chart does *not* suggest. *It does not suggest any effect of Arabic forms of the dedication, either actual or hypothetical, on the origin of the Latin forms. Rather, The Latin and Arabic forms seem to descend and develop, independently, from a common archetype or original.*

3. The Manuscript Titles, and the Forms of Incipits and Explicits

Most of the manuscripts have an Incipit which gives either a translation of the Greek word *pneumatika* or some other indication of the contents of the book; some manuscripts have an Explicit which restates or varies this indication or medieval book title.

The titles differ in specificity. Some are highly particularized, so for example the "Pneumatic Devices for Water Works," *De spiritualibus ingeniis ducendi aquam*, used in **R.** Others have much more general expressions, such as "Surprise Engines," "Subtle Devices," "Admirable Devices," or simply "Devices," which in Greek are called *thaumastà* or *thaumantia* (a term also used in a contemporary, Peripatetic work[138]); in Latin they are called *De ingeniis subtilibus*, **C, G;** *Super . . . mirabilibus ingeniis . . .*, **K;** or simply *De ingeniis*, **X.** Some of

[137] Corrected to "marzaton."

[138] The Peripatetic *Mechanical Problems*, p. 848a34–37.

Chart 3. Stemma of Dedications in Philo Manuscripts

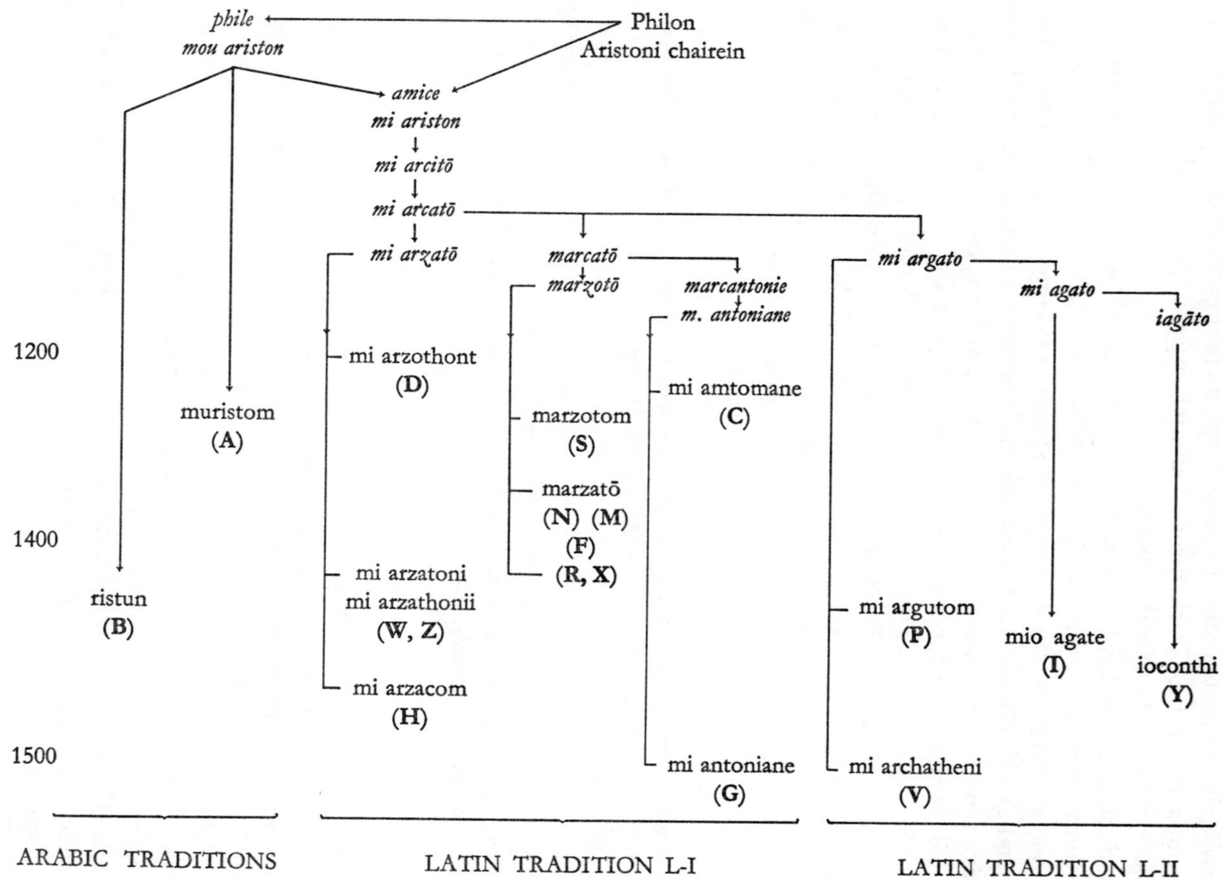

Chart 4. Titles of *Pneumatica* Manuscripts (Archetypes, Copies and Arabic Versions)

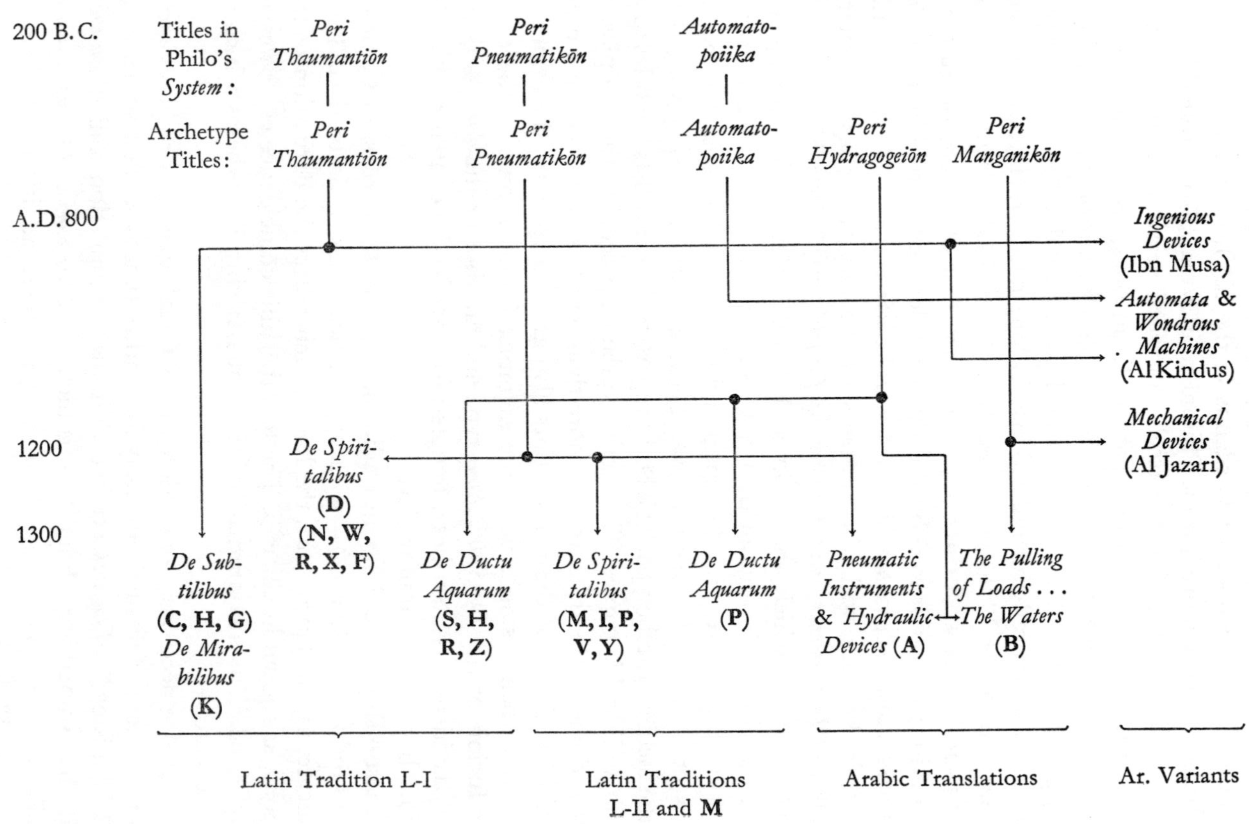

these forms, for example *De [ingeniis] subtilibus*, reappear in later works, such as the well-known book of Cardanus in the XVI. century.

Elements of the more specific Philonian titles may also be found in the Arabic Philo tradition, including the freely compiled versions of Ibn Musa and his successors. The Arabic titles, like the Latin ones, use different selections of title elements anciently attributed to Philo's *Pneumatica* and related books. For example the copy of the Arabic translation, contained in manuscript **A**, begins with a specifically formulated inscription, in the first line of the cover page, saying: "Book of Philo on *Pneumatic Instruments and Hydraulic Devices*." The next page begins with three short lines, the first of which contains the ritual invocation; it is followed by the further title: "Book of Philo on *Pneumatic Apparatus and [Hydraulic Devices such as] Cups and Ewers*." At the end of the book, in keeping with the contents of the last two chapters, reference is made to "*Instruments and Machines for [Furnishing] Water*."

The way in which the Latin and Arabic titles originated can be reconstructed with a fair amount of detail, as shown in Chart 4. This chart begins, in its upper left part, with the reported titles of books contained in the pertinent parts of Philo's *Mechanical System*. These and other titles appear, thereafter, in the medieval archetype. The bottom part of the chart lists the titles of existing Latin and Arabic copies. The Arabic variants, preceding these copies, are also listed. Some of the titles appear more than once, as they occur in a number of traditions. Some of the manuscripts also are shown more than once, as they use more than a single title. For example, **P**—as shown in the chart—has one Incipit of the *De Spiritalibus* type and another of the *De Ductu Aquarum* type. Evidently the scribes attempted to describe the contents with some precision, but without uniformity.

Chart 4 shows the different traditions—except the earliest or Celtic-Carolingian-Gothic one, Chart 2—with hypothetical lines of derivation. These lines coincide, in substance, with the lines which reflect the derivation of the dedications, as shown in Chart 3. The fact that this coincidence can be established may be taken as a confirmation of the underlying hypotheses which are presented here.

Chart 4 reflects the conclusion that a Greek copyist or abbreviator—evidently in late Hellenistic or early Byzantine times—added something about "Mechanisms," *Manganika* to the Philonian compilation, called *Pneumatica*,[139] and that something about "Aqueducts," *Hydragogeia*, *Ductus aquarum* was similarly added. The relationship between these additions and the reported books of Philo is unclear.

[139] Schmidt on Carra de Vaux, pp. 327 ff.

An original title referring to pneumatics, *Peri pneumatikōn*, seems to reappear in part **A** of the Arabic tradition, and to be generally translated in Latin as *Spiritualia (Spiritalia)*. The latter term is often abbreviated to *Sp'alia*. The abbreviation has sometimes been read and reproduced as *Specialia*, but such reading was erroneous. It is clear from Philo's authentic announcement of the book on pneumatics that the reading *Spiritualia* is the correct one. For good measure this is confirmed by **W**, where the first folio of the volume indicates the contents as *Alkimia et Philo de spiritualibus ingeniis*; the Philonian treatise then follows (fo. 92v) under the title *De sp'alibus ingeniis*.[140]

Chart 4 disregards lost manuscripts, such as the reported one of Leonardo.[141] It also disregards minor differences between titles, for example the difference between *De conductibus aquarum* (**Z**, Incipit) and *De aquarum ductibus* (**Z**, Explicit).

In effect, this survey shows that three titles, *Thaumantia*, *Pneumatica* and *Hydraulica*, appear in Latin and Arabic forms and can be attributed to the archetype, and that two further titles—*Automata* and *Manganika*—appear only in the Arabic forms and can be attributed only to the eastern prototype.

None of the Latin titles reflects the particular selection, adopted in any of the five Arabic documents.

V. Chapter Numbers, Chapter Titles or Introductions, and their Origins

The Arabic Philo, manuscript **A**, designates all drawings by a system of figure numbers. As substantially each figure corresponds to a single chapter, these designations can be used as a system of chapter numbers. Carra de Vaux used almost the same system; he only tried to reconcile the numbers with a set introduced by Schmidt, which assigned separate chapter numbers to some unillustrated text portions. For this reason the modern numbers of Carra de Vaux generally are about 3 or 4 units higher than the authentic, Arabic ones. In spite of this defect, the present edition adopts the modernized number system of Carra de Vaux, in order to simplify comparisons. It adopts this system comprehensively, and disregards the variations of Rose, Schmidt and others.

In manuscript **A**, the chapter identified by modern number 38 refers to a Chapter IX. On review it appears that Chapter number IX was a prototype

---

140 So also **F** and **W**. In **N**, the white leather binding has a small, brownish slip pasted to the middle of its top edge portion, with this notation in Gothic writing: "*Opus alfragani: Messalach de motu coeli: linconiensis[?] de fractionibus radiorum: philonius de specialibus ingeniis Et plura naturalia et medicinalia.*"

141 L. Reti[78], p. 81. The recorded title was: "*De aquis.*"

number corresponding to modern Chapter 13. As rightly pointed out, or at least generally suggested, by Carra de Vaux, the Arabic statement indicates an early origin and good preservation of at least the first two scores of chapters. By detailed correlation of old and new numbers and identified contents, in Chart 5, we can, accordingly reconstruct the chapter numbers of the Arabic prototype. Their general sequence coincides with an authentic Latin one, but *the actual numbers, in the Latin manuscripts, are different.*

The difference is shown by two copies belonging to the tradition L-II: the Paris and Egerton manuscripts. These, exceptionally, include fragments of a Latin chapter number system. (The copies of tradition L-I and the others of L-II do not show chapter numbers. Only one of the late copies, **G,** has a somewhat similar system. It uses letters, ranging from "a" to "n", for Chapters 1–12 and 17–21. This copy also has a non-standard arrangement of the drawings).

Chapter titles, in a modern sense, are not used in either the Latin or the Arabic manuscripts. There are only indirect indications of the chapter divisions. Substantially each chapter, both Latin and Arabic, has one illustration. In substantially all copies, as also in the variant of Ibn Musa, the drawing is said to follow, and actually follows, at the end of a chapter.

Additionally, the beginning of a new chapter or group of chapters is shown in many cases by an initial, or at least by leaving room for an initial, or by the use of a special, calligraphic line. These lines and initials are frequently done in color. A typical calligraphic line occurs in Palatinus Latinus 1332 (**W**), ahead of the chapter now numbered 5. It reads: *De natura aque.*

Still further, the divisions of the book are indicated by the use of introductory formulas. In the Latin Philo and in the first score of Arabic chapters, beginnings of chapters are indicated by transitional phrases of some significance for the functional descriptions, for example: "What I will explain now is similar to what I said about . . ., but . . ." In some cases the phrase shrinks to the mere words "Ad idem" or "Item." The last two scores of Arabic chapters differ, as they use, instead, laudatory statements of no technical significance, for example: "Let us now describe another elegant vessel," or: "Construction of another surprising vessel."

By comparison with other pneumatics books, left by Hero, Ibn Musa, and the Latin copyists, we may be able to come to a view of the way in which the structure of the archetype came about. In rough outline we can see vestiges of early books or divisions of Philo's larger work contained in the archetype. These early divisions (Pneumatics, Automata, "Aqueducts") are shown in Chart 5. As was indicated in Chart 4, they are also reflected by titles of existing manuscripts. In some sections of the manuscripts, for example from Chapter

22 to 39, we can see evidently illogical positions of chapters, as we also do in Hero's book of pneumatics. The impression arises that the writer of an archetype, prototype or copy sometimes began to copy or abbreviate a new chapter, regardless of its topic, while the ink was drying on his copy of another chapter, forming part of another group of chapters in the source document. The resulting confusion is now reflected in the Philonian and Heronian manuscripts.

The reference from Chapter 38 back to Chapter IX, in manuscript **A**, allows us to reconstruct chapter numbers of the Arabic prototype. This in turn enables us to reconstruct a table of contents of this prototype. The reconstruction is shown in Chart 5.

Chart 5. Reconstructed Table of Contents of the Arabic Prototype, Reflected by Authentic Chapter Numbers in Arabic Manuscript **A**

	Prototype Chapters Reflected in **A***	Modern Chapter Numbers
PNEUMATICS		
Introduction	(I)	1, 2
Basic Pneumatic Theory	(II, III), IV, V	3/4a, 5/6, 7, 8
Siphons and Clepsydras	(VI), VII, (VIII)	9, 10, 11/12
Siphon Combinations	IX, X, XI, XII	13, 14, 15, 16
Flooded Vent Clepsydras	(XIII), XIV, XV, XVI	17, 18, 19, 20
Clepsydra Combination	XVII	21
Interrupted Siphon	(XVIII)	4b**
VARIANTS AND AUTOMATA		
Pneumatic Variants	XIX–XXVI	22–29
Dispensers	XXVII–XXXIII	30–36
Further Pneumatic Variants	XXXIV–XXXVI	37–39
Float Automata	XXXVII–XXXIX	40–42
Wine-Thieves & Jacket Vessels	XL–XLVII	43–50
Combination Faucets	XLVIII–LI	51–54
Fire Automata	LII–LIV	55–57
Acoustic Automata	LV–LXI	58–63
"AQUEDUCTS"		
Hydraulic Power Machines	LXII, (LXIII)	64, 65

* The numbers shown in parentheses are interpolated. Those without parentheses are translated from **A**.

** Later displaced to the present position, between 4a and 5. Also see note 447, about possible, earlier positions of 4b.

The date of this Arabic prototype is unrecorded, but internal evidence indicates that the date was subsequent to the early 1200's: one of the major chapters, copied after 1300 in manuscript **A** (Chapter 30), is derived from the work of AlJazari, which is dated 1205.[142] It is interesting to note that *the Latin prototype already was being copied or re-copied, in those early 1200*'s, as we will see when reviewing the early Latin copy **D**.

VI. Scope of the Philo Manuscripts and of the Volumes Containing them

1. Scope of the Philo Manuscripts themselves

We can now move from the formal elements to the real contents of the work. We can begin by trying to determine the extent to which the chapters of one tradition appear in the other. The pertinent facts are presented in Chart 6, which is developed from Charts 3 and 5 and from a few other, documented facts.

Chart 6 begins with Chapters 1–12, which contain the basic demonstrations of pneumatic theory according to Philo. In view of the announcement, at the end of the war engine book, that a book on pneumatics would follow, and also in view of the structure of most of the existing manuscripts, which begin with at least an attempt to set forth pneumatic theory, such theory may be taken as a principal object, if not the only object, of the original Pneumatica.

This principal object is well, if not ideally, ser forth in in the Latin tradition, but is covered poorly in Arabic copy **A:** a *lacuna* begins in Chapter 4 and continues throughout Chapter 5 and far into 6. The Arabic Chapters 9–12 are also full of *lacunae* and displacements. Some of these difficulties, but hardly all of them, may be chargeable to the copyist of manuscript **A,** who generally appears to have been a careful and responsible scribe. The basic, theoretical part of the science is not covered at all in the Arabic extracts, collected in manuscript **B.** Nor is this part even mentioned by Ahmed Ibn Musa, or by AlJazari, or by any documented part of the works of AlKindus. The writers of those eastern works show an almost exclusive interest in the demonstration of assorted contraptions, either pneumatic or mechanical, pneumatic, or pyrotechnical. A similar interest is shown by the Gothic fragments. Briefly: *theoretical pneumatics is fully shown in the Latin translation, not in the Arabic translation or variants or the Gothic extracts.* By contrast, the various contraptions and assorted techniques are shown much more fully in the Arabic tradition.

---

142 Carra de Vaux, p. 161 n, correctly states this origin of Chapter 30. For further details see Wiedemann-Hauser on AlJazari, Drinking Vessels.

Details of the copying process are unknown, and so are the absolute dates for the archetype and prototype documents, but some relative dates of such documents are probable, also in view of the development of dedications and titles. *It hardly seems possible that L-I, L-II and the early, Latin-Gothic tradition could simply be translated, with or without "basmalah", from a single Arabic prototype which can be documented in any form. It also hardly seems that the Latin prototype, underlying the Celtic or Carolingian or Gothic fragments, can have a date as late as that of the Arabic prototype, considered in the preceding section of these notes.*

Rather, the characteristics of the eastern and western traditions, considered here, suggest largely independent work, done by different groups of translators

Chart 6. Basic Traditions of Chapters 1–65

The Byzantine Archetype as Reflected in Manuscript **A**		The Gothic Tradition	Latin Tradition L–I	Latin Tradition L–II
[I, II][a]	= Ch. 1–4a	lacuna	1–4a	1–4a
[III]	lacuna	,,	5, 6	5, 6[g]
IV, V	7, 8	,,	7, 8	7, 8
[VI], VII	lacuna	,,	9, 10	9, 10
[VIII]	11, 12[b]	,,	11, 12	11, 12
IX–XII	13–16	,,	(13–16[d])	lacuna
[XIII]	17	,,	17	17
XIV–XVII	18–21	,,	18–21	21, 18–20
[XVIII]	lacuna	,,	4b[e]	4b[h]
XIX, XX	22, 23	,,	(22, 23[f])	lacuna
[XXI . . .]–XXXII	24–35	,,	lacuna	,,
XXXIII–XXXIX	36–42	,,	,,	,,
XL–LI	43–54	,,	,,	,,
LII–LIV	55–57	55–57[c]	,,	,,
LV . . .–[LXIII]	58–65	lacuna	,,	,,

[a] Bold cursive type: probably from the *Pneumatica* of Philo and his Hellenistic pupils. Plain cursive type: probably from Philo's *Automatopoiica*. Plain roman type: probably from later, Persian works. Brackets: the numbers are not legible in **A**.

[b] Displaced in **A**.

[c] Reflected in **C, H,** Villard, and Kyeser.

[d] Reflected by drawings in **D** and partly in **N** and **Z**.

[e] Displaced in L–I.

[f] Reflected by drawings in **D**.

[g] L–II more detailed than L–I.

[h] Displaced.

and illustrators, and differently derived from the Greek source documents. Some of these groups may have derived their information from more than one source—just as Hero had done in his *Automatopoiica*. Such consultation of different sources may explain the strange cross-influences, shown by some existing documents in their invocations, and by others in their illustrations.

2. Scope of the Manuscript Volumes:

Contents other than the *Pneumatica* Copies[143]

The Philo copies, other than **A,** were bound into volumes—sometimes called "Scientific *Miscellanea*", in the manuscript collections—, which also contain copies of other works. Many of these compile copies of famous works or chapters on astronomy, optics and general physics, as well as metaphysics, with the more practical booklets on geometry and surveying, chemistry, dye-making, metal work, and minting, to list only some of the most popular topics.

It is not intended in this edition of Philo's work to explore these compilations in depth, but some remarks on them are called for. They show a climate in which Philo's work was preserved. The work was not read in a vacuum. It was read as part of the post-Alexandrian handbook literature, along with works of higher, natural science and of course along with the ubiquitous works of spiritual edification: these were the intellectual intake of workers in the scriptoria, where the manuscripts were copied and then bound in the *miscellanea* volumes, and to some extent these also were the intellectual intake of artisans.

A similar climate may have led to the compilation of pneumatic and non-pneumatic works, including Philonian and non-Philonian works, in the archetype periods.

Interesting among the works compiled with Philo are those on atmospheric phenomena, such as the *Presages of Storms*, in **S** and elsewhere. **H, I** and **K** have tracts on sources for water supplies.[144] The late humanist copy, **V** combines Philo with Ptolemy and Euclid and, perhaps significantly, with the small Arabic-Latin treatise *De crepusculis et nubium ascensionibus*, which for the

[143] Some pertinent details may be found in writings of medievalists. For example, Thorndike, III, pp. 400ff., 441ff., 473ff., 741ff. describes pertinent manuscript copies, mainly of works of Oresme.

[144] Derived from Vitruvius, Book VIII, through unknown medieval intermediaries. In **H,** the water-finding tract is on fo. 144r, v; in **I,** on fo. 9–11; and in **K,** on fo. 104v–105v. Variants attributed to Martini are in the edition of his *Trattati*, Martini[3], pp. 160–170 and 355–359.

first time since Aristotle discussed, and approximately established, the height of the atmosphere.[145]

Two manuscripts combine their copies of Philo translations with their tracts on sources for water supplies (**H** and **K**), whereas one manuscript (**I**) presents this latter tract separately, after its Philo version. One of the combining manuscripts, **H,** attributes the water tract to a Magister Longobardus; the other, **K,** attributes similar chapters to Landfridus; the chapters are also said to be derived from the Heronian "De inani et vacuo." At least substantial parts of the water tract come from the VIII. Book of Vitruvius.

A student of one of our manuscripts has criticized this combination as "make-shift."[146] Indeed it is hard to find a plausible rationale for it. Nevertheless the criticism seems too sharp. In these manuscripts, and similarly in **X** and A. S. 2755, the compiler transmitted works of ancient science which plausibly appeared to him to be inter-related, as they, to some extent, mutually supplemented one another. The scribes of earlier times, who had appended fragments of a book on "aqueducts" to Philo's "pneumatics", had done substantially the same thing. It appears that Greek archetypes of Philo's *Pneumatica* as well as Latin and Arabic prototypes derived therefrom, enriched our knowledge of early science by preserving sets of Alexandrian and post-Alexandrian, Philonian and pseudo-Philonian writings, about various sciences and technologies, and by making their individual selections from such sets.

VII. Style and Contents in the Manuscript Copies

1. The Illustrations

a. Their Position and Size Relative to the Text

In the Arabic copies **A** and **B,** each illustration of the first two scores of chapters, and almost each of the last score, is inserted directly below the text describing it. Depending on the position of the last written line, the drawing follows in the middle or in the bottom of the page or on top of the next page. A similar arrangement is used in the Latin copy **Z.** In this arrangement a drawing occupies, typically, about one quarter to three quarters of a page, or sometimes a full page.

Somewhat more complex is the system used in the earlier Latin copies **D, M** and **W,** wherein this arrangement occurs only in a few of the illustrated

145 The possible significance has been mentioned in the historical discussion of Galileo, Torricelli, and Kircher.

146 Birkenmajer[102], pp. 23–26, 30 ("Machwerk").

chapters; others have their illustrations in the side margins, or in large, otherwise blank areas of the sheets.

A few Latin manuscripts (**C, P, Y**) and the Italian translation (**I**) have the drawings inserted in rectangular areas, left free for this purpose by the text. This system is also used at a few points near the beginning of the last score of drawings in **A** (fo. 59b, 63a anp b). In one manuscript using this system (**C**) the drawings occupy very small parts of a page; they are inserted in the corresponding text portions and fill only part of their width, after the manner of initials and small-scale miniatures.

Still another tradition places all drawings in the side and bottom margins (so mainly **N,** also **F, K, L, S, V**).

In **G,** all drawings are on a separate sheet, mentioned, after the Explicit of the text, in a distichon conceived in Renaissance Latin:

> *Hoc folium extrahas, tu qui cognoscere formas*
> *Supradictorum, lector, cupis ingeniorum.*

The book-binder overlooked the fact that the sheet belongs to the Philo manuscript, and inserted several quires of paper and text between them. In similar ways, the use of separate "atlas" sheets or "plates" for the drawing may have caused their loss in four other manuscripts, **E, H, R,** and **X.**

The first arrangement of drawings, represented by **A, B** and **Z** reflects the structure of the ancient "*book-roll*"; the last few, represented by **C, S,** and **G,** use the principal styles of the "*codex*" arrangement.[147] Some of the systems, mainly **D, M** and **W** and perhaps **Y, P** and **I,** may be said to be intermediates. It is interesting that all Arabic documents and a few Latin ones belong to the roll style, whereas the majority of the Latin ones belong to the codex style, but that *none of the roll-style documents in Latin is any closer to the Arabic documents than to the other Latin ones, in the details of execution and style.* We come to these details.

b. Drawing Techniques

Elementary drafting tools—straight-edge and compass—are used in **A, S, V** and **Z.** No use of parallel rulers is in evidence. No such tools at all are used in **N** and in many other copies.

---

[147] About these problems, see Weitzmann[67], *loc. cit.*; same, *Illustrations in Roll and Codex*, Princeton, 1947; and same, *Ancient Book Illustration*, Harvard, 1959. The latter two works are cited here, respectively, as "Weitzmann, 1947" (mainly for copies of papyrus rolls, Figures 35–53) and "Weitzmann, 1959."

The rules of perspective drawing are unknown throughout these illustrations. Only at one or two points, in Chapters 30, 35, the eastern draftsman makes a small, intuitive, quasi-perspective attempt. No such attempt is reflected by any of the western copies.

The copyists do not indicate dimensions, on the drawings of Philonian devices. Only a few medieval layouts of Heronian constructions, which are inserted at the end of **G** and of **X,** are dimensioned.

Various colors are used in the manuscripts. For example the Philonian drawings are done in red in **C,** and in green and yellow in **S.**

c. Drawing Styles

In the drawing copies of western origin, the devices are shown almost entirely without ornamentation. By contrast, many elements of decorative style occur in the illustrations of **A,** executed by an Arabic artist. Among other things, this artist draws charming human and animal figures to animate some of the drawings. The sources of his style are difficult to define, and thus far no attempt has been made to trace them, as Carra de Vaux omitted the authentic pictures from his edition, and replaced them with an insipid exercise of the *art nouveau* of his time. The following are a few pertinent suggestions. For them I am heavily indebted to Gustina Scaglia.

Arabic Chapters 33–36 include figure motifs which may perhaps be assigned to the heavily stylized "Celtic" or "Anglo-Irish" type, best known from the famous Books of Durrow, of Lindisfarne, and of Kells (c. 700–800). This style has parallels in other parts of northwestern Europe, but hardly anywhere else. It is most surprising to see it reflected in an Arabic manuscript. No figures of similar kind occur in the book of Ibn Musa, or in other parts of the eastern tradition, such as manuscript **B.**

The Arabic illustrations reflect a variety of eastern drawing styles or traditions, mainly by the contrast between Chapters 59, 61–65 on the one hand and most of the preceding ones on the other.

They also reflect another influence in their representation of oddly distorted Corinthian column capitals, mainly in Chapter 18 and less pronouncedly in Chapters 33–35. Earlier reflections of such capitals occur in Carolingian manuscripts and architecture. Byzantines may have transmitted the strange motifs of the illustrations to these chapters, both to and from Carolingian artists.[148] The distorted column capitals also appear in medieval copies of Hellenistic drawings; one of these in turn is reproduced in the publication of 1693, shown here as Figure 1.

---

[148] Weitzmann[67], pp. 67, 75f.; Hartwig[126], p. 186n.

d. Contents of the Drawings

In some chapters, neither the draftsmen nor the scribes appear to have understood the proposal of Philo, or of the author of the archetype. Such is the case particularly in Chapter 4b, where parts of the text are incomprehensible and the drawing shows an inoperable device. Even here, the draftsmen tried, albeit unsuccessfully, to make sense of the text, and it is remarkable how close at least the draftsman in **C** came to a meaningful sketch. He may have understood the technical requirements of the device more clearly than the authors of the prototype or archetype understood them. It is even conceivable that he understood these requirements more clearly than Philo did.

Elsewhere, the draftsmen are unable to clarify important details, left to them by the writers or copyists of the written description. Such is the case in Chapters 33, 34 of **A,** where perspective or developed uses of plan view and elevation would be needed.

Some of the illustrations do not comply with the text. For example the texts of Chapters 2 and 6, 8 and 17 describe certain jugs as narrow-mouthed, and the Latin drawings so show them; **A** shows all of them as wide-mouthed. In Chapter 7 the Arabic text calls for a "cup"; the drawing shows a glass with an ornamental, twisted stem. Partition walls, described for Chapter 21, are recognizable (although with difficulty) in some of the Latin variants; they are totally confused in **A.** Rotary valve parts, described or suggested for Chapter 22, are freely rendered in **D** but are totally unrecognizable in **A.** *These are further elements which make it hard to believe, with the current theory, that the Latin Philo descends from an Arabic one.*

A similar difficulty arises elsewhere. The spout tubes of "*ibriq*" vessels, in Arabic Chapters 13, 23 and 24, have generally plain, downwardly curving ends; in the Latin variants they have straight, ornamented ends. The base of the oil lamp in Arabic Chapter 20 is flat; it is in form of an animal-footed tripod in a few western copies, such as **M.** These distinctions may come from the use of differently styled utensils, not from differences of graphic style, in different scriptoria. In any event they show antiquity closer to the Latin manuscript **M** than to the Arabic copy **A.**

Briefly, the existing western illustrations appear more abstract, slightly more old, and probably more authentic, as well as evidently less intuitive and less innovative than their eastern parallels.

2. Types of Expression in the Philo Texts

There are four types of expression in the Philo texts. First, in the introductory part of Chapter 1, stilted and involved language is used in the Latin Philo, and the terms are copied in such different forms, in the various manuscripts, as to indicate that they puzzled the copyists.[149] In second place, we have the balance of the Latin text; in third place the corresponding Arabic texts; and in fourth place the Greek fragments, preserved as Philo's pneumatic remarks in his *Belopoiica* and as Hero's copy from Philo's *Automatopoiica*. These three are simple and easy; they differ only in the languages that they use.

a. The Latin Introduction

The first few sentences in Chapter 1 of the book read:

Dixit:[150] Quia tuum, amice . . ., iam novi desiderium ad sciendum ingenia subtilia,[151] voluntati tue obnoxius,[152] interrogationi sedeo.[153] Petitioni [tue de] hoc libro[154] componendo libens[155] acquiesco. Presentem siquidem diligentie tue[156] instituo tractatum, [ut] in[157] omnibus, que in[158] huiusmodi quesieris ingeniis, tibi non incommodum sit exemplum. De his itaque tractare incipiens, imprimis assero, quod harum scientie rerum[159] nulli sapientum[160] nature sunt

---

149 About this kind of style, see the histories of Greek rhetoric under the heading "Asianism;" for example Pauly-Wissowa *s.v.* Hegesias of Magnesia (in Philo's time).

150 **S, W, Z** omit "Dixit." **X** writes "Dixerunt et." **Y** originally had "Dixit" but corrected it to "Dixerunt et." **H:** "Dixit Philonius."

151 **S:** "Satis novi . . . subtilia ingenia." **X** omits "ad sciendum." **Z:** "aliquid sciendi ingenia spiritualia," corrected to ". . . ingenia subtilia."

152 **F:** obnossius.

153 **H:** "Interrogationi sive petitioni . . ." **C, W:** "interrogationi respondebo." **D, Y:** ". . . respondebo." **G, P, S:** "r[espo]ndeo." **V, Z:** ". . . tue respondeo." The unusual word "sedeo" is confirmed by **F, X** and, in the form "interrogationem sed[e]o," by **R.**

154 **V** omits "tue."

155 **D:** libenter. **H:** libris *(sic)*.

156 **X** omits "tue".

157 **D:** "ut in." **H, P, S, Y:** "ut de." **F:** "de." corrected to "in". All copies use the medieval form "ut . . . non . . . sit," at this point and many others. Sometimes they use classical Latin; for example, further along in Chapter 1: "ne . . . prolixitatem . . . incurrat."

158 **D, W, Z:** "que de . . .;" **S** omits "in."

159 In **R, S, X** and **Z** the clause "imprimis . . . rerum" is partly illegible. In **V** and **Y** it is omitted. **D, W:** "harum materie rerum . . ."

160 **D, W:** "non nulli sapientum." **R:** "nulli sapientum sunt non . . ." **H:** "nullo nulli." There is no basis for Rose's reading "nonnullis," which was questioned but accepted by Schmidt. No copy has "*sapientium.*"

incognite.[161] Unde[162] et philosophi, qui in[163] rebus consideraverunt naturalibus, plurimorum[164] erronee opinioni non consentientes, dixerunt . . .

Are these Latin formulas derived from a lost Arabic prototype, or do they reflect Greek "Asianismus," translated in the Latin prototype? A conclusive answer seems impossible at this point. Surely the language is medieval, when it, for example, uses the form "*ut . . . tibi non incommodum sit.*" The language also is unusual, when it, for example, uses the form "*interrogationi sedeo,*" (a form which may be derived from some use of the Greek "*heō*" rather than from an Arabic clause).

In the collation of Latin forms, for the introductory sentences, I have taken little account of variant readings which only reflect minor preferences of spelling and the like, such as "*obnossius*" instead of "*obnoxius.*" It appears from a comparison of the manuscripts that the Latin scribes gave free rein to their own preferences, in minor matters of this kind, and that they freely "corrected" the prototype spelling.

The collation accounts for substantially all other variants. It shows that these arise from attempts to eliminate obscurities of the prototype, rather than from giving rein to individual preference. In other words, the copyists made a strong attempt, in the introductory sentences as well as in the invocation and dedication, to reproduce the authentic message. For example, substantially all versions retain the prototype's poetic form "*sapientum*", which had been adopted instead of the regular "*sapientium.*"

The obscurities of the introduction are substantial. Did the author refer to his friend's desire in five different forms (*desiderium*, *voluntas*, *interrogatio*, *petitio*, *diligenti*)? If so, why and how? Did he mean that "the pertinent sciences are known to all learned men" (*scientie harum rerum sunt nature incognite nulli sapientum*), or that they are known "to all natural philosophers" (*. . . sunt incognite nulli sapientum nature*)? Are these statements attributable to Philo? They have no close parallel in his Greek writings on military engineering. However, since these are abbreviations of later times, it is conceivable that an authentic, Greek introduction, filled with Asianisms, happened to remain unabbreviated. The Latin variants indicate that the scribes in the XIII. to XVI. centuries were puzzled by this introductory text, and tried to clarify it.

b. The Technical Descriptions in Greek

We still have two precious fragments, in Greek, of Philo's writing on pneumatics and automata. One of them occurs in his *Belopoiica* and describes

[161] **S:** "sunt non incognite," perhaps based on a misreading of "sunt n'e incognite."

[162] **S** omits "Unde."

[163] **P:** "de".

[164] **H:** "Plurimorum eorum." **Y** omits "consideraverunt."

the Ctesibian experiments with a compressed air device for a catapult; the other is provided by Hero—by far the most qualified reporter of the last two millennia—and describes, with acknowledged use of Philo's book, an automaton which displayed the "Legend of Nauplius."[165]

These fragments of Philo's technical Greek, dealing with topics included in the present book, will be recorded, in substantially full translation, in one of the following notes on the western Philo traditions. They are most helpful for the evaluation of other Philonian or Pseudo-Philonian texts in the *Pneumatica:* they provide by far the most authentic documentation of Philo's pertinent style. Texts which resemble them are probably authentic; those which do not are probably spurious. This theory applies regardless of whether the texts are otherwise elegant, or inspired, or prefaced with some *initia.*

c. The Descriptions of Pneumatics, in Latin and Arabic

Beginning with Philo's reference to the earlier philosophers *(philosophi dixerunt . . .)*, and throughout the remainder of his book, the text of the Latin translation dispenses with the initial, formalistic, and repetitive statements. It uses simple and straightforward recitals of facts, and of conclusions suggested by them.

The Arabic descriptions usually are longer than the Latin ones, and are substantially more affected with formalism and repetition. They also are more filled up with basically irrelevant, quantitative remarks, including many remarks of this meaningless type: the vessel can have two or four or any desired number of volume units.

Neither the Latin nor the Arabic translation creates the impression in general, that the translator found great trouble in the technology to be expressed, or was aware of actual problems, offered by the devices; however, subsequent copyists may have misunderstood translations correctly rendered in the prototypes, as is indicated by the fact that the different, Latin copies differ quite materially in their renditions of the Latin translation. They are rather uniform only in that they are much shorter and simpler, in stating each proposition, than the renditions of the Latin introduction. An example may be seen in a collated transcript of a few sentences from the Latin Chapter 2:

---

[165] The Greek texts of both are most conveniently shown by *Veteres mathematici*; although some details have been clarified since then, for example by Diels-Schramm, the errors of the Paris edition have been exaggerated.

Accipiatur vas, quale predocui, non amplo ore, in cuius fundo[166] sit foramen minutum, et obstruatur cera. Deinde convertatur vas, ore verso,[167] et ponatur in profundo aque... Quod si paulatim et suaviter extrahatur, invenietur interius siccum, in nulla ipsius parte preter os exterius madefactum.[168]

These statements are almost identical in all Latin texts. They also are nearly identical in the Arabic translation. The latter translation differs only in that it speaks a second time of an "Egyptian" vessel, when the Latin version refers to the "aforedescribed" vessel:

Let us take an Egyptian carafe with a narrow mouth. Let its bottom have a minute hole, and let it be closed with wax. Then let it be turned over, and let it be put in deep water . . . and then taken out slowly and gently. The inside will be found dry. It is wetted only on the mouthpiece.

Empedocles had described substantially the same demonstration, performed with a clepsydra sieve. Anaxagoras had elaborated on it. Aristotle had mentioned it briefly and polemically. Philo repeats the report of these predecessors and agrees with its substance (at this point). The Latin as well as the Arabic translators clearly understand the physical meaning of the demonstration.

The subsequent operation, described by Empedocles, also elaborated by Anaxagoras, had been disregarded in the pertinent statement of Aristotle, but was fully described—although not necessarily explained—by Philo in Chapter 11. The translators again seem to uniformly understand Philo's description, as reported in the archetype. They are uniformly unaware of the interpretation given by Empedocles; it remains an open question whether, or to what extent, Philo was aware of that interpretation.

In the manuscripts of Cambridge and Geneva, **C** and **G,** the scribes appear to seek clarification of some expressions, by making the text more explicit. They do this uniformly, and thereby indicate that the later manuscript **G** depends on **C.** For example, both refer, in Chapter 1, to "ingenia subtilia *et arcana*," and to "scribendo librum *nobilem;*" in Chapter 2, both change the word "invenietur" to "*reperietur*," "ore" to "*orificio*," and "cera" to "*cum* cera." They also share the strange dedication to "Antoniane" or "Amtomane."

The technical text of the Latin prototype, which already is quite laconic, is subjected to further, drastic abbreviation in the abstract of **E,** and in the extract of **M.**

---

166 **P:** . . . quale predocuimus, ampliare, in cuius fundo . . . *(sic)*. **H** changes "fundo" to "profundo". **S** inverts "in cuius" and omits "cera," at the end of the sentence.

167 **S:** "Tunc dehinc convertatur . . .;" **C:** "ore scilicet stricto verso."

168 In **W,** "interius siccum" is omitted. Here and in **X,** eight words are inverted.

In humanistic times, still another kind of change is encountered, mainly in **K,** where the scribe largely rewrites the text. He hardly introduces a scientific or technical innovation, but attempts to improve the Latin. He also tries to make the drawings more realistic. He begins Chapter 2 as follows:

Aliqua amphora vel phiola non perforata, cuius corpus sit latum, collum vero longum et strictum, si mergitur aque violenter, ita quod phiale sint sub aquis et fundum supra, numquam intrabit aqua in ampullam, nisi prius evacuetur aer occupans locum in phiala. Ergo est corpus. Aliter: dicta phiala sit profunda in aqua. Si paulatim et suave extrahetur totum interius ampulle apparebit siccum, os autem exterius madefactum. Quod non esset nisi aer esset corpus.

There are only few distinctions of substantive nature, between the different manuscripts. They relate only to details. The relatively strongest distinction has to do with the water-lifting candle of Chapter 8. In most manuscripts, water-lifting is said to occur in this device because of the natural rising of the fire: *propter ignis ascensionem*. In **A** and **V,** by contrast, it is attributed to the burning of the fire: *propter ignis accensionem*, and its air-destroying effect (which is mentioned also in the other manuscripts).

The manuscripts differ more radically with regard to the style of abbreviations. For example the following differently abbreviated statements appear in five Latin manuscripts, at the head of Chapter 8:

Q' mō dtūrus sū p'eunti silā' . . . (**N**)
Qd' m̊ dictur' sum p'cedenti si'at' . . . (**D**)
qmō d'turus sū p'eūto sīliat' . . . (**X**)
quod modo d't'r' sum p'cedenti assimilat' . . . (**S**)
Quod v'o modo dictu' sū p'cedenti p't assimilari . . . (**C**)

In a few cases, analysis of the abbreviations will be proposed, hereafter, in an attempt to establish a sequence of Latin copies.

In rare instances the Latin texts show errors about technical details.[169] Such errors are more frequent in the Arabic texts. They usually occur at the points where the Arabic translator introduced pseudo-accuracies of really fictitious character.[170] *The text translations, particularly of Chapters 11, 12 and 17–21, show that the Arabic errors do not reappear in the Latin text.*

In both translations, the Latin and the Arabic one, each chapter usually begins with a general description of the device and of its function; it then introduces reference letters; and in the end explains how the referenced parts pro-

[169] The principal errors in substance occur in Chapter 6, where **P** says: elevabitur ad locum ymum." **N** says: "descendens ad locum altum." The error in **P** is corrected.

[170] Cf. Hinz[103], Vorwort.

duce the result.[171] In this respect, Philo proceeds after the manner of Euclidian "proposition, construction, and conclusion," which can also be said to be the method developed by classic rhetoric for scientific demonstrations. The "proposition" statements usually are corrupted in the copybooks. In the Arabic copies they are usually dissolved into laudatory statements. In the Latin copies a full-scale "proposition" occurs only in Chapters 2, 8 and 10; elsewhere a short, introductory statement is used, as already mentioned in connection with chapter titles.

d. Summary as to the Technical Descriptions and Translations

A frankly personal evaluation follows, recording the editor's summary of the facts, shown up to here, regarding the manuscript texts.

Throughout the *Pneumatica* manuscripts, the formulations are uneven in form and quality, although the contents of the chapters are closely interrelated. The first score of chapters, in both Arabic and Latin formulations, uses an effective teaching technique for explaining simple propositions of natural law by simple demonstrations, in the field of pneumatics. Chapters 37 to 39 belong to the same field, and are of the same character.

In some other parts of the book the demonstration is equally effective, but is less didactic, and does not deal with pneumatics. This group contains mainly automata (Chapters 33, 34, 36, 40, 41 and 42). Those automata may be authentically Philonian, or may belong to the Persian devices also described by Ahmed Ibn Musa, or may partake of both characteristics.

Still another group comprises the wine-thieves and the combination faucets, described in subsequent chapters, and other automata and machines. These have little or nothing to do with "pneumatic" technique; their descriptions differ from the preceding ones in that they are expressed in much more naive terms, regarding technical fundamentals and they appear to be much affected by the regression of technical understanding which characterizes the dark ages of the pneumatic sciences. Incidentally, some of these other chapters include explicit statements which allude to contrasts between the earlier constructions and the "new" ones: Chapters 32, 56, 57 and 63.

Therefore I incline to consider the contents of Chapters 1–21 and 37–39 as reflecting the remaining contents of Philo's original book on pneumatics. The initial and theoretical part of the book is reflected more fully and therefore more adequately by the Latin Chapters 1–12 than by the Asian parallel. The Arabic automata Chapters 33, 34, 36, 40, 41 and 42 reflect the remaining contents of Philo's original book on automata, while other automata chapters are largely, if not entirely, based on interpolations, inserted in late Hellenistic times,

---

171 Drachmann, p. 43.

which are partly reflected also by the Gothic fragments. These other chapters are characterized not only by showing no interest in theoretical physics, but also by indicating little talent for didactic exposition. However, they reflect superior mechanical techniques.

Some of the devices described and illustrated in the manuscripts of the Pneumatica are inoperable. This is true of Chapters 4b, 28, 43 to 45, and 64. In the case of Chapter 4b the problem may be due either to the loss of details, incident to the shifting of the original chapter, or to insufficiency of Philo's original disclosure. In all other cases the trouble seems to be due to late Hellenistic, Persian or Arabic interpolations, as is clearly the case in the illustrations of the Ibn Musa variants.

Assuming that Philo's style, in his original *Pneumatica*, had similarity with the existing Greek abbreviation of his *Belopoiica* and with Hero's quotation from his *Automatopoiica*, we may reconcile such style most easily with the one that is used in the Latin Chapters 17 to 20 and the Arabic Chapters 40 to 42. It progresses leisurely but logically from element to element, with uniform reference to each element and without repetition of statements. It does not seek rigorous causations or proofs, but suggests plausible interconnections. In other chapters this kind of language seems more or les corrupted: by dwelling on details and insertion of euologies, in the Arabic version, and by excessive abbreviation of the technical exposition, in the several Lation versions, including the Gothic fragments.

So far as the Latin and Arabic copies share complete chapters (that is, combined descriptions and illustrations), they are fairly uniform, in both translations, as to their basic contents. In their correlation of theoretical physics with simple demonstration, they are quite superior to the roughly contemporary copies of the books of Ahmed Ibn Musa and AlJazari.

VIII. Marginal Sketches and Glosses

A few non-typical drawings occur in the Latin manuscripts. For example **Z** has one of them, associated with Chapter 6. The drawing may be intended to show a length of flexible siphon hose, rolled up at both ends.

S has two sketches and a pertinent note, which appear to suggest clarified versions of Chapter 10. The note suggests a new "experimentum" with a vessel having a tall central structure or "calamus" surrounding another similar structure; the sketch resembles a corresponding Heronian drawing (I, 13). Other Heronian drawings reappear in **G** and **X;** a Ptolemean instrument appears in **K;** and unexplained constructions appear in **D** and, similarly, in **W.** In ms. **A,** an unexplained drawing is among those combined with Chapter 35.

Finally, there a few unillustrated, marginal remarks or glosses. One of them may be seen in **D,** Chapter 11; another occurs in **N** and is transcribed in this edition (Chapter 13). These are written by the scribes. By contrast, in **R,** Chap. 1 and 8, a reader draws attention to the fact that the vacuum is discussed in the text. In **W** someone writes in the margin: "Explicat in primo capitulo quod aer est corpus." Evidently these latter glosses may stem from scholastic readers of the book. In general they, as well as the marginal sketches, give the impression that the contents were a matter of positive, physical interest, not only of historic concern.

IX. Reference Letters

A fragment of the Greek *Pneumatica*, itself, has survived, in form of the use of a certain sequence of reference letters. The archtype evidently applied letters of the Greek alphabet to the elements of drawings, mentioned in the descriptions. The medieval followers, who translated these letters into Arabic and Latin ones,[172] almost invariably left them in their Greek sequence, when introducing them into each chapter's description. *They usually begin this sequence with the Greek "a, b, g, d, e, z;" rarely with the Latin "a, b, c, d, e, f;" and never with the Arabic "a, b, t, ṭ, g."*[173] The complete system of letters, as used in their Philo translations, can be shown as follows.

α	β	γ	δ	ε	ζ	η	ϑ	ι	κ	λ	μ	ν	ξ	ο	π	ρ	σ	τ	υ	ου
a	b	g	d	e	z	h	th	i	k	l	m	n	x	o	p	r	s	t	y	u
ا	ب	ج	د	ه	ز	ح	ط	ع	ك	ل	م	ن	[–]	ه	پ	ر	س	ت	ع	و

Some details: The last four letters of the Greek alphabet (phi, chi, psi, omega) do not occur in the Philo chapters. The first letter of Philo's name, in the book title, becomes "f" in Arabic and either "ph" or "f" in Latin. In the drawings and descriptions, *epsilon* and *omicron* become "e" and "o" in Latin; both become "*tan*" in Arabic but the omicron sometimes appears as "u". The western manuscripts, in contrast to the eastern ones, occasionally add the Latin letter "q" or the Latin subsequence "xy", and often change "g" to "c". The eastern manuscripts sometimes interpret the Arabic letters as numerals.[174]

Detailed comparison between Arabic and Latin references is possible in seven chapters: numbers 7, 11 and 17–21. (Such comparison is otherwise

---

[172] With rather strange, Graeco-Latin forms mainly for the Latin "g," for example in **I** and **S.**

[173] The history of these sequences is described by H. Jensen, *Die Schrift in Vergangenheit und Gegenwart*, Berlin, 1958, pp. 302–308, 424f., and 480.

[174] Particularly in Chapter 39. For the successive forms of "Arabic" numerals: F. Cajori, *A History of Mathematical Notations*, Chicago, 1928, I, pp. 49–51.

impossible, because of the Arabic and Latin *lacunae* and because there are no reference letters in Chapters 1–3, 5, 6 and 12). In each of the seven chapters with references, the reference letter system is the backbone of the technical teaching, as had been the case, before Philo, in the works of Euclid and in some of those of Aristotle.[175] In comparison with these earlier works, the Arabic Philo shows uncertainty in the interpretation and use of its reference letters. *The Latin Philo uses its own selection of Greek letters, and uses it with greater assurance and effect.*

For example: In Chapter 17, the Arabic translator applies the first six letters to certain parts; in Chapters 18–21 he seems to have forgotten this tradition, although the same parts recur. The Latin translator retains his own selection of basic letters, throughout the set of Chapters 17–21.

The Arabic edition remains inaccurate, with respect to its use of reference letters, also in parts of the manuscript which have no Latin parallel. For example: in Chapters 33, 37, 42, 56 and 64, some of the Arabic, textual reference letters are poorly presented in the drawings; in Chapters 21, 22, 29, 38, reference letters of the drawings are altered or disregarded in the text. Are these shortcomings to be charged to the Arabic prototype translator-draftsman, or to the draftsman of the existing manuscript? It seems more probable that they are to be charged to the prototype; the copywork, shown by the manuscript, seems more accurate.

It will be understood that the Arabic tradition had ample compensation, even if it was inaccurate with regard to some details such as the preservation of the Greek reference system. The accuracy of a "copy" is significant for the historical reconstruction of the past, but the originality of a "variant" has equal significance for the renewed progress. Both processes, recovery and renewal, are slow and difficult. We may be pleased that "variants" as well as "copies" of Philo's seminal work were produced. I have tried to outline the historic progress from archaic times to Democritus, Strato, Ctesibius, Philo, Hero, and the renewed progress to Ibn Musa, Aben Moat and Torricelli. The present, documentary notes were intended only to clarify the relative significance of the Latin copying effort, applied to Philo's text book, for the transition from the dark ages of scientific indifference and retrogression to the age of renewed, eastern and western progress.[176]

---

[175] *Physica*, VI, 1 and 2, pp. 231b18 to 233b31. About the reconstruction of the Aristotelian drawings: Weitzmann[147], 1959, pp. 8–10; M.S. Schramm, *Die Bedeutung der Bewegungslehre des Aristoteles*, Frankfurt, 1962, pp. 11–60.

[176] A.C. Crombie, *Augustine to Galileo*, London, 1961, II, p. 38n, believes that Philo's *Pneumatica* was not widely known in Scholastic times. However, it was more widely known than Hero's *Pneumatica*, even if Scholastic writers cared little about either.

The Western Traditions of Philo's *Pneumatica*

Facsimile of the Philo manuscript in CLM 534

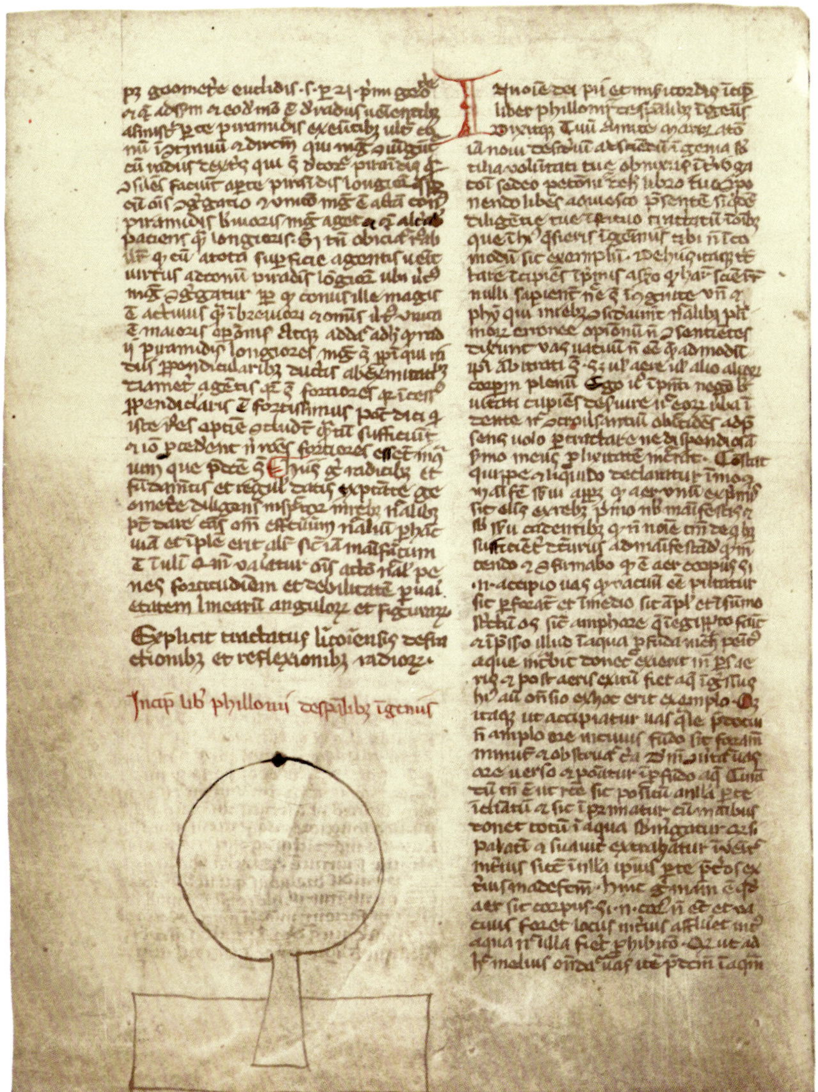

Figure 3. CLM 534, fo. 26v. Chapters 1 and 2; Drawing of Chapter 2.

Figure 4. CLM 534, fo. 27r. Chapters 2 (end), 3 and 4; Drawing of Chapter 4.

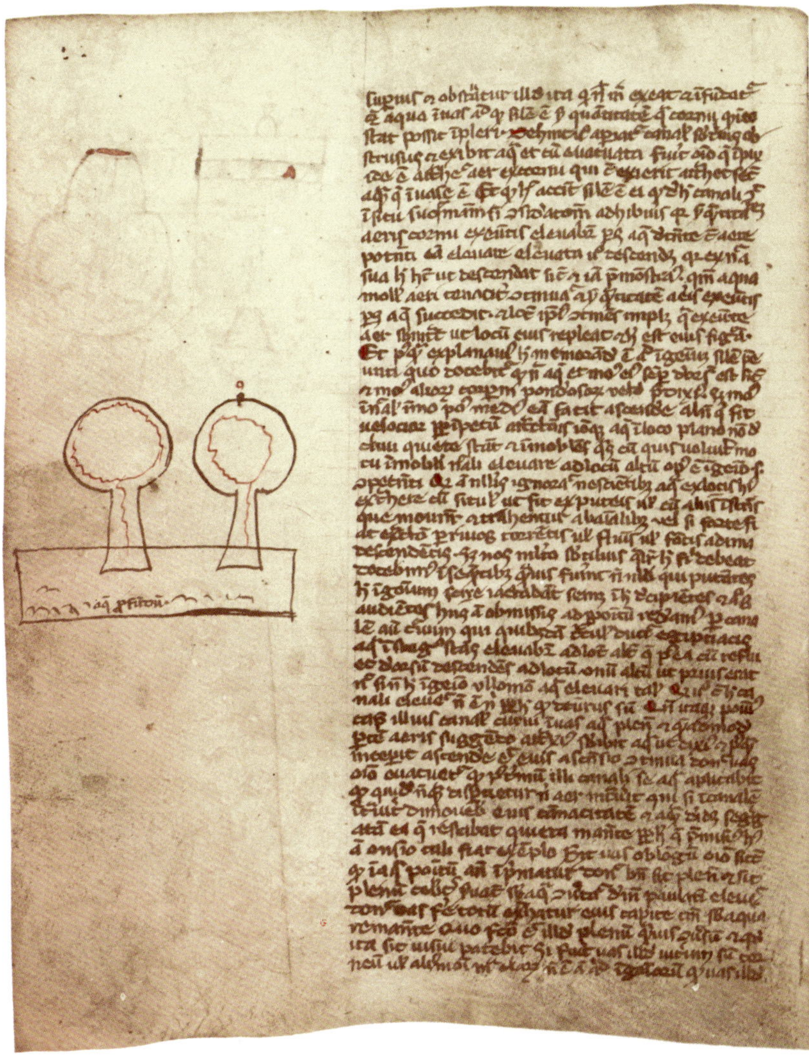

Figure 5. CLM 534, fo. 27v. Chapters 4 (end), 5 and 6; Drawing of Chapter 6.

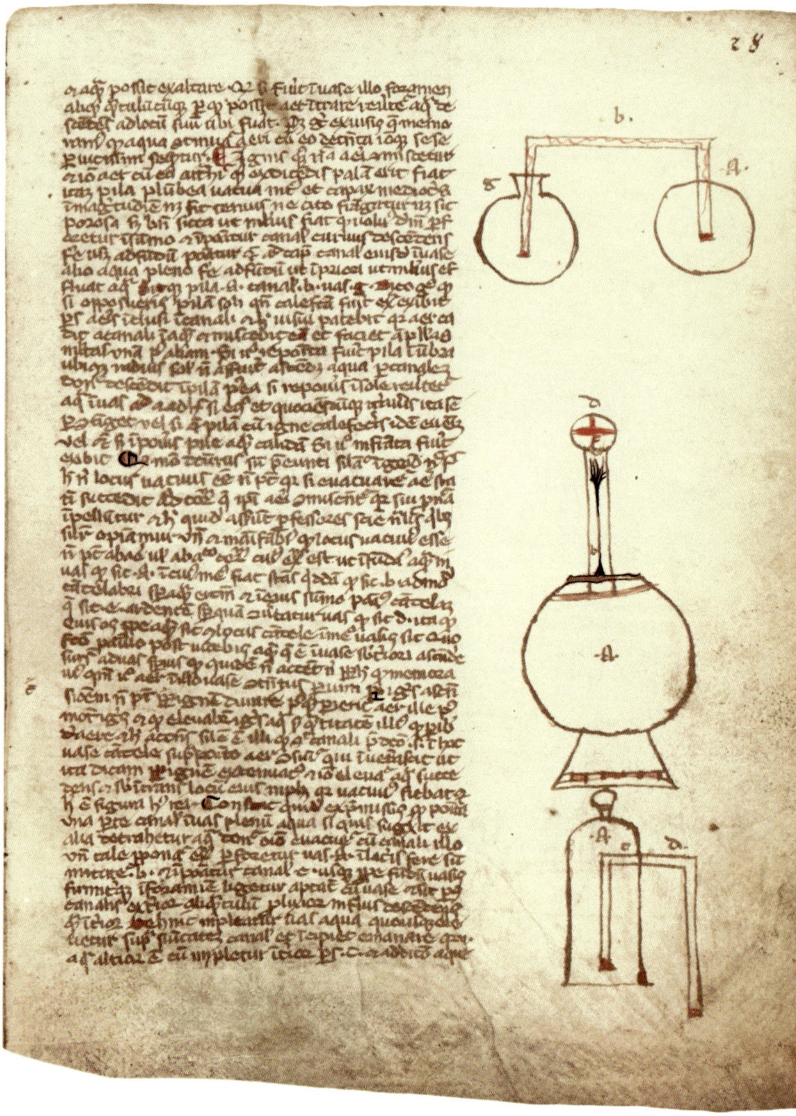

Figure 6. CLM 534, fo. 28r. Chapters 6 (end), 7, 8 and 9;
Drawings of Chapters 7, 8 and 9.

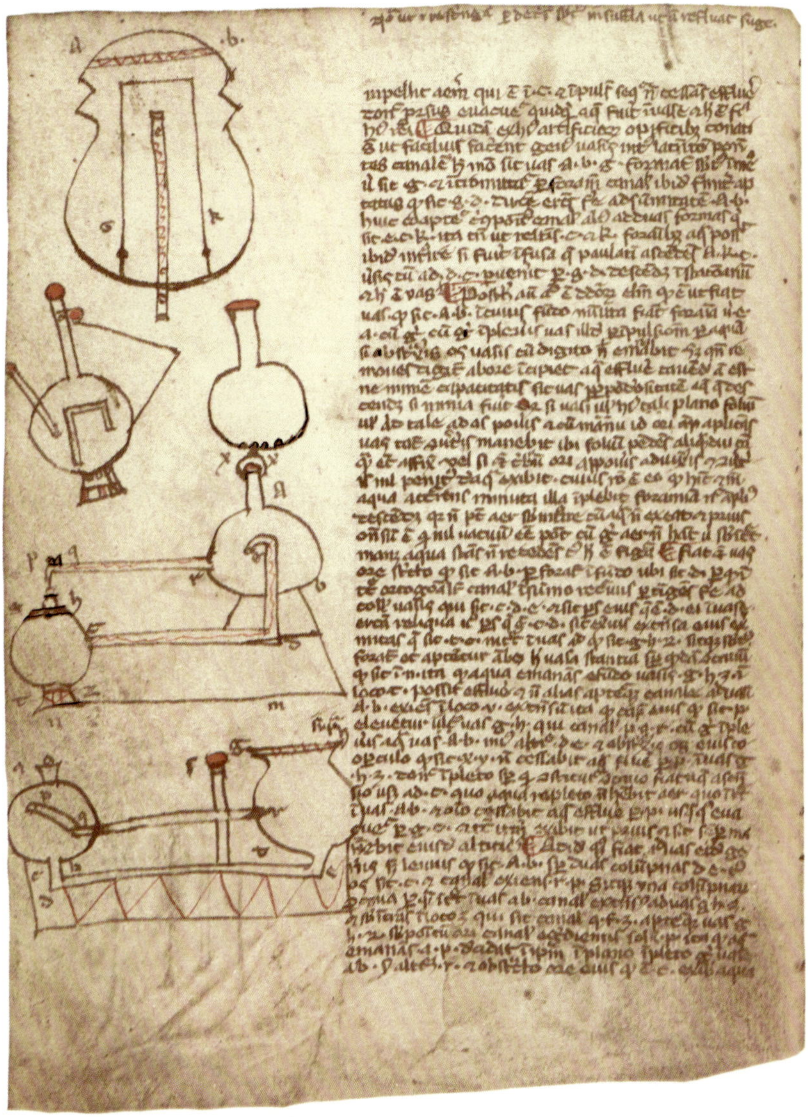

Figure 7. CLM 534, fo. 28v. Chapters 9 (end), 10, 11, 12, 13, 17 and 18; Drawings of Chapters 10, 11, 13, 17 and 19. Text of Chapter 13 is added at the top.

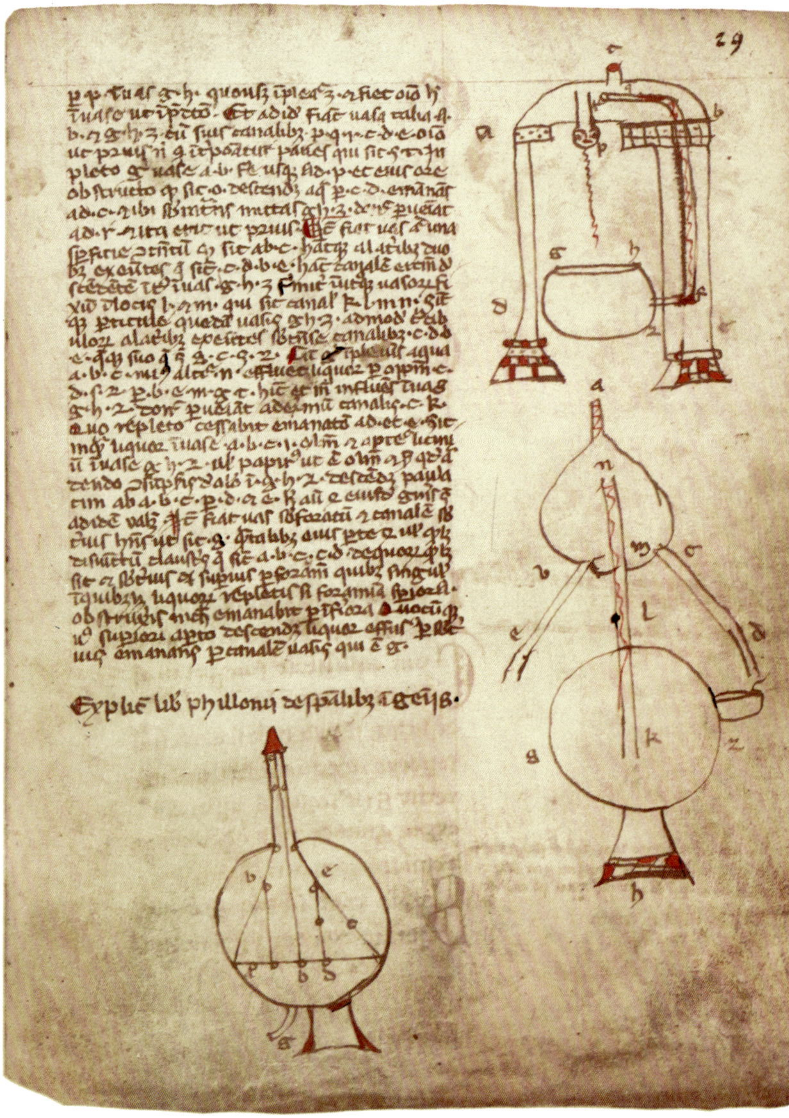

Figure 8. CLM 534, fo. 29r. Chapters 18 (end), 20 and 21;
Drawings of Chapters 18, 20 and 21.

Collated Transcript of the Latin Philo

A new transcript of the Latin Philo, as recorded in CLM 534 (**N**), follows here. In footnotes I have collated **N** with the score of other Latin copies, known to me at this time, and primarily with the Cambridge copy, which frequently and intelligently amplifies the prototype expressions. I have given much attention also to other early texts, such as Digby and Sloane, which contain problematic expressions, some of which reflect apparent problems of the Latin prototype.

Generally, the consensus of the Latin codices agrees with CLM 534. However, exceptions have been noted, also in the later copies.

It is the purpose of this collated transcript to come closer to the construction of a Latin standard text, and thereby to the reconstruction of a Latin prototype. Variants are noted mainly insofar as they affect the scientific-technical message. The collation generally disregards details of only philological interest, such as variant spellings, abbreviations, and sequences of words. For information on such details, the reader must consult the facsimile pages or the manuscripts themselves. The transcript adds a system of modern punctuation. It is compared with the Arabic tradition, in notes to the translation of A. S. 3713, which follows later in this edition.

Incipit liber Phillonis *de sp[iritu]alibus ingeniis.*

In nomine dei pii et misericordis incipit liber Phillonis *de sp[iritu]alibus ingeniis.*[201]

[Chapter 1][202]

Dixit: q[uia] tuum, amice marzatom,[203] iam novi desiderium ad sciendum ingenia subtilia, voluntati tue obnoxius, interrogationi sedeo. Petitioni tue [de] hoc libro componendo libens acquiesco. Presentem siquidem diligentie tue instituo tractatum, [ut] in omnibus que in huiusmodi quesieris ingeniis, tibi non incommodum sit exemplum. De his itaque tractare incipiens, imprimis assero, quod harum sciencie rerum nulli sapientum nature sunt incognite.

---

201 Other forms of name, invocation and title: see the documentary introduction, text to notes 124–141.

202 Variant readings in Chapter 1: notes 150–164.

203 Other forms of dedication: Chart 3.

Unde et philosophi, qui in rebus consideraverunt naturalibus, plurimorum erronee opinioni non consentientes, dixerunt vas vacuum non esse, quemadmodum ipsi arbitrati sunt, sed vel aere vel alio aliquo corporum[204] plenum. Ego vero, in presenti negotio brevitati cupiens deservire, nec eorum verba in dente[205] nec controversantium objectiones volo pertractare, ne dispensiosam sermo meus prolixitatem incurrat. Constat quippe et liquido declaratur, immo et manifeste sensui apparet, quod aer unum ex primis sit el[ement]is, ex rebus primo nobis manifestis et sub sensu cadentibus quidem, non nomine tantum. De quibus sufficienter dicturus [sum] ad manifestandum quod intendo et confirmabo, quod est aer corpus.

[Chapter 2]

Si enim accepero vas, quod vacuum esse putatur — sit perforatum[206] et in medio sit amplum et in summo strictum os, sicut amphore que in Egipto fiunt — et impressero illud in aqua profunda, nihil penitus aque intrabit donec exierit inde pars aeris. Et post aeris exitum fiet aque ingressus. Huius autem ostensio ex hoc erit exemplo.

Oportet itaque ut accipiatur vas, quale predocui, non amplo ore, in cuius fundo sit foramen minutum, et obstruatur cera. Deinde convertatur vas, ore verso,[207] et ponatur in profundo aque. Curandum tamen est, ut recte stet positum, a nulla parte inclinatum. Et sic imprimatur cum manibus, donec totum in aqua submergatur. Quod si paulatim et suaviter extrahatur, invenietur interius siccum, in nulla ipsius parte preter os exterius madefactum.

Hinc igitur manifestum est quod aer sit corpus.[208] Si enim corpus non esset et vacuus foret locus interius, accederet intus aqua nec ulla fie[re]t prohibitio.

Quod ut adhuc melius ostendatur, vas iterum predictum in aquam demergatur, ut prius, et cera super foramen posita abstrahatur. Quo facto erit exitus

---

[204] **D:** aliquo alio. **R, V:** alio aliquo corpore. **S:** aliquo aliorum corporum. **Z:** vel alio vel aliquo aut aere.

[205] **D, F, P, V, W, Y, Z** read "inducere" instead of "in dente." **G** changes it to "introducere" and **S** to "int[er]serere." The reading of **N** is fully confirmed by **R,** "in dente," and also by **X,** "intente;" it seems to reflect the Latin prototype's reading most exactly. The only meaningful reading is that of **C:** "intendo." Schmidt accepts the reading of **P,** without discussion of **S** and **N.** The present collation will use such silent selection only at points where it seems certain that the selected text is meaningful and that the variants are minor.

[206] **C:** in hunc modum formatum. **D, W, Z** etc.: sic formatum.

[207] Variant readings for this sentence and the next three: above, n. 166–168.

[208] The radically abbreviated copies, **E** and **M,** omit the rest of this chapter.

aeris per foramen, sensui manifestus in ampullis aque si fuerit aqua super foramen. Et implebitur vas propter exitum aeris per foramen.[209]

Et quod facit aerem necessario [exire] est modus [motus] et impetus aque impellentis, prius impulse in vasis intromissione. Et hec est demonstratio quod aer sit corpus.

[Chapter 3]

Declarabo etiam naturam aliorum verborum,[210] que ad id sunt utilia, quod inquiritur in hac scientia.

Quidam autem sapientum fatebantur aerem ex minutissimis constare corporibus minimisque particulis, que propter suam parvitatem sub sensu visus vel aliquo aliorum non cadunt, quando segregata fuerint, nisi fallantur.[211] Cum autem coeunt simul coniuncta, non est ita.

Dixit quoque unus ex sapientibus,[212] quod inane sit horum natura et commisceatur corpori aeris, cum eo videlicet,[213] quod fuerit in suis partibus minutis subtilibus, sicut etiam ex partibus[214] omnis rei mollis, immo ex omnibus rebus.

Et iam declaravi hoc alias sufficienter, unde feci mentionem *de arbitriis mirabilibus*.[215] Premissum autem persequamur exemplum.

S[u]b[st]an[ti]a igitur elementi humidi detur aeri continuari, prius quam alii nature commixtione tenaci, et non est inter ea distancia.

Ideoque contingit multotiens aque, quod eius sit iter sic, quod cum aere elevatur. Si vero natura sue ponderositatis,[216] humoris,[217] vincens fuerit super illam, que modica est, impellit eam deorsum, ut descendat deorsum, quia et omnium corporum ponderosorum hec est proprietas et consuetudo, ut ex natura sua deorsum descendant.

---

[209] **D, W,** etc.: exituram . . .; **S:** . . . per aquam. NOTE: Here and elsewhere the use of ellipsis (. . .), in this collation, indicates that the other words are identical, or substantially identical, with those of **N.**

[210] **D, Z:** [D]eclarabo . . . motum aliorum elementorum. **R** (as in note 205 and elsewhere) confirms **N**'s reading: "verborum." **C:** [d]Eclarabo . . . melius aliorum elementorum. **W:** Eclarabo naturam modo aliorum . . .; **V:** . . . nihilominus modum . . .

[211] **C, V:** quando segregata fuerint non fallaciter. **D, W, Z:** . . . nisi fallaciter. **R** also reads: . . . nisi fallantur.

[212] **C:** antiquis sapientibus: one of the explicit amplifications of **C.**

[213] **D, W, Z:** Cum ea videlicet.

[214] **D:** sic etiam partibus. **W:** sic in partibus.

[215] **C, D, F, G, S:** ubi feci mentionem . . .; **Y, Z:** arbitriis mutabilibus.

[216] **D:** natura sua. **W:** natura sue ponderositatis.

[217] **C, R, S, V, W, Z** omit "humoris." In the following clause, **D, W, Z** omit or modify "que modica est." The commas around "humoris" of course are editorial.

[Chapter 4]

[a]

Quod autem aqua sursum multotiens elevatur, manifestum est. Attrahitur enim cum aere elevato, quia et continuatur.[218] Sicut patet in hoc fit in vase,[219] cum quo gustatur vinum. Cuius capud cum quis in ore tenuerit, sugendo aerem qui in eo est, attrahitur[220] atque cum aere corpus molle liquidissimum, scilicet quod subtus est, quia aeri continuatur, tanquam ei cum visco applicaretur, vel aliusmodi ligamento.[221]

[b]

Quod etiam sic manifestabitur. Oportet utique [itaque], ut sit sumatum cornu bovinum, intus bene accavatum, donec amplum sit et tenue et sincerum. Sit autem altitudinis mediocris et eius forma ut forma pinee rotunda et pineata.[222] Et ex parte qua assimilatur figure pineali, ei subterius[223] applicetur vas ligneum, valde siccum, ut bene fiat quod volumus. Ita vero coaptetur, ut in nulla eius parte aer possit expirare. Et sic erit cornu quantitas ad modum et similitudinem pixidis.

Tunc convertatur hoc cornu, stans in vase quodam quod ore sit amplum, ut est cyphus.

Item ponetur et alia pixis, plumbea, subtus vel iuxta,[224] bene aptata, ut, cum opus fuerit, nichil inde exeat aeris. Super ambas quoque has pixides aptabitur canalis, valde siccus, capita habens recurva et deorsum ad fundum utriusque pixidis pertingentia. Habebitque pixis plumbea os modicum, desuper paulum erectum, ut aqua cum vase aliquo per illud possit infundi. Fiet quoque in hac pixide subterius canalis alius, brevis et modicus, ut inde evacuetur aqua, quando voluerimus.

---

[218] **C:** cum aere et elevatur quia . . .

[219] **D:** quod contingit de [vase] scilicet vineo. **R, S, V, Z:** quod contingit in vase. **W:** quod contingit de . . . cum [quo] gustatur (omitting quo by apparent oversight; leaving room after de.)

[220] **C, S, V, Z:** attrahit. **W:** attrahit scilicet vinum.

[221] **K** "shortens" Chapter 4b to a single sentence: "Si quis ore ad os canne vel phiale in [cu]curbite stante vel huiusmodi, aerem attraxerit sugendo . . . [atque?] aquam . . .

[222] **D** omits "ut forma . . . et pineata" and marks the lacuna by dots. **W** omits it without remark.

[223] **V:** subtilius.

[224] **W:** alia pixis subtus plumbea vel iuxta. Most manuscripts have "vel," however, see note 344, 416.

Exempli causa sit cornu nota "a", vasi simili cypho [nota "b"], nota pixidis plumbea "g", et canalis illius parvi nota "d", et canalis magni communis nota "e",[225] et oris parvi quod supra pixidem est plumbeam, est nota "z".

Quibus omnibus secundum quod docui coaptatis, constituatur [obstruatur] canalis modicus subterius et impleatur plumbea pixis aqua, per os superius, et obstruatur illud, quod nihil inde exeat. Et infundatur etiam aqua in vas aliud,[226] quod [cypho] simile est, secundum quantitatem, qua cornu, quod in eo stat, possit impleri. Dehinc vero aperiatur canalis subterius[227] obstructus et exibit aqua. Et cum evacuata fuerit omnino, que in pixide est, attrahetur aer ex cornu. Qui cum exierit, attrahet secum aquam, que in vase est. Et quod hic accidit, simile est ei, quod de hoc canali contingit. In situ suo[228] manifestum [est], si considerationem adhibueris, quod secundum quantitatem aeris, cornu exeuntis, elevabitur pars aque detente cum aere, potenti eam elevare. Elevata vero descendet, quia ex natura sua hoc habet, ut descendat, sicut etiam premonstravimus. Quoniam aqua mollis aeri tenaciter continuatur et secundum quantitatem aeris exeuntis, pars aque succedit et locum ipsius continens implet. Qua exeunte, aer subintrat, ut locum eius repleat. Et hec est eius figura.[229]

[Chapter 5]

Et postquam explanavimus hoc, memorandum est aliud ingenium, simile preeunti, quo docebitur,[230] quod natura aque et motus eius semper deorsum est, sicut et motus aliorum corporum ponderosorum, velud prediximus, sed motus innaturalis,[231] immo pocius medius, eam facit ascendere aliquando. Qui fit velocior propter impetum attraccionis.[232] Ideoque aque in loco plano et non declivi quiete stant et immobiles. Quas cum quis voluerit motu [innaturali][233] elevare ad locum altum, opus est ingenio sibi competenti.

---

225 The text of **N** shows "d" cancelled by underdotting and followed by "e".

226 **V:** vas illud.

227 **D, W:** superius. **P** omits the word.

228 **C, W:** visui tuo. **D:** visui tunc; similarly **F, G, H, S, Z,** and others.

229 **C, S, Z** and others: huius rei figura (or: . . . demonstratio, or: . . . signum, or: figura que sequitur). **D** omits this clause. **W** has a chapter title at the beginning of Chapter 5: De natura aque. **K** omits Chapter 5.

230 **S:** docebatur. Evidently a mistake of the copyist.

231 **S:** "naturalis," another plain error of the copyist.

232 **S:** abstrac[ti]onis. **W:** sue attractionis. **Z:** scilicet attractionis.

233 **C, D, W, Z:** motu innaturali. **N:** Motu immobili naturali. **S:** motu naturali. It seems that the prototype used a confusing abbreviation for "innaturali."

Quod a nonnullis ignoratur, nescientibus[234] aquam ex locis his extrahere [nisi][235] cum situlis, ut fit ex puteis, vel cum aliis instrumentis, que moventur et trahentur ab animalibus, vel si forte fiat extractio per rivos torrentis vel fluminis vel fontis, ad ima descendentis. Sed nos multo subtilius, qualiter hoc fieri debeat, docebimus in sequentibus, quamvis fuerint nonnulli, qui, putantes hoc ingenium scire, iactabant [se librum compositures, quod penitus ignorabant, et][236] semet in hoc decipientes et alios audientes. His autem obmissis, ad propositum redeamus.[237]

[Chapter 6]

[a]

Per canalem autem curvum, qui quibusdam circulus[238] dicitur egiptiacus, aqua in stagno stans elevabitur ad locum altum, que postea autem refluet deorsum, descendens ad locum [imum],[239] ut prius erat. Nec sine hoc, ingenio [predicto], poterit aqua ullo modo elevari.[240] Quod vero cum hoc canali elevetur,[241] non est nisi propter hec, que dicturus sum.

Quando itaque posuerimus capud illius canalis curvi in vas aqua plenum et quandam admodum partem aeris, sugendo,[242] attraximus, subibit aqua, ut diximus. Et postquam inceperit ascendere, erit ascensio continua, donec vas omnino evacuetur, quod postremum illi canali se aqua applicabit.

Que quidem nunquam dispercietur nisi aer intervenerit. Qui si in canalem intravenerit, dimovebit eius tenacitatem, et aquam dividet segregatam, ea, que restabat,[243] quieta manente, propter hoc quod premisimus.

---

[234] **C:** sapientibus nescientibus: "who as [mere] philosophers do not know."

[235] **D, W, Z,** etc.: "nisi," as interpolated here.

[236] **C, D, W, Z:** "se librum . . . ignorabant," as interpolated here.

[237] **W** has a block diagram opposite Chapter 5. An upper block says: "Aque descensus, . . . innaturalis, tardus, velox." A middle block: "Medius tardior naturali, velocior naturali." A lower block: "Idem . . . parte eius ascensui . . ." The words omitted here are illegible.

[238] **C** omits circulus. **D, W, Z:** circinus. **V:** cyrrius. **K:** Fiat una canalis de ligno molli vel lapide vel tota vitricata . . . *(sic)*.

[239] Corrected in accord with **D, W, Z.** The text in **N** says: descendens ad locum unum altum. **C:** . . . ad locum unum.

[240] See the translation. **N** and others omit "predicto."

[241] **V:** quod non cum hoc canali elevatur *(sic)*.

[242] **C:** suggerendo. **D, W, Z:** cum ore sugendo.

[243] **S:** ea restabit quiete. **V:** eaque restabit quiete. **C:** ea que restabit.

[b]

Huius autem ostensio tali fiat exemplo. Sit vas oblongum, omnino siccum, quod in aqua positum autem imprimatur, donec bene sit plenum, et sic plenum servatum, celeriter sub aqua convertatur. Deinde paulatim elevetur, donec vas fere totum extrahatur, eius capite tamen sub aqua remanente. Quo facto erit vas illud plenum quamvis conversum, et quod ita sit, visui patebit, si fuerit vas illud vitreum sive corneum vel alius modi materia clarum. Non est autem aliquid ingeniorum quo vas illud aquam possit [exalare],[244] quam si fuerit in vase illo foramen aliquod quantulumcumque, per quod possit aer intrare. Revertetur aqua descendens ad locum suum, ubi fuerat.

Patet igitur ex universis que memoravimus, quod aqua continuatur aeri, cum eo distenta. Ideoque se semper vicissim sequuntur.[245]

[Chapter 7]

Ignis quoque natura aeri commiscetur et immo aer cum eo attrahitur. Quod ex dicendis palam erit. Fiat itaque pila plumbea, vacua intus et capax,[246] mediocris in magnitudine, nec sit tenuis, ne cito frangatur, nec sit porosa,[247] sed bene sicca, ut melius fiat, quod volumus. Deinde perforetur in summo et imponatur canalis curvus descendens fere usque ad fundum. Ponatur etiam aliud capud canalis eiusdem in vase alio, aqua pleno, fere ad fundum, ut in priori, ut melius effluat aqua.[248] Sitque pila "a", canalis "b", vas "g".

Dico igitur quod si opposueris pilam soli, quando calefacta fuerit, extra exibit pars aeris inclusi in canali. Et hoc visui patebit, quia aer cadit a canali in aquam et miscebit eam et faciet ampullas multas, unam post aliam. Si vero reposita fuerit pila in umbra, ubicumque radius solis non affuerit, ascendit aqua per canalem, donec descendit in pilam.

Post ea, si reposueris in sole, revertetur aqua in vas aliud. Et adhuc si e converso et quotiescunque iteraveris, ita semper continget. Vel si etiam pilam cum igne calefeceris, idem eveniet, vel si imposueris pile aquam calidam. Si vero infrigidata fuerit, exibit.[249]

---

[244] So mainly **C, G, P, S,** also **I.** The text in **N,** and also in **D, F, V, W, X, Z** uses the confusing word "exaltare." On the other hand, **C** and **S** confusingly change "quam" to "quod."

[245] **C:** cum eo descendente. Ideoque se sequuntur *hec duo elementa.*

[246] **C:** in[t]us vacua, *stricto orificio.* **K:** Ignis proprietas est calida, per quam aeri commiscatur . . . Fiat una pila de plumbo concava et capax et intus vacua . . .

[247] **K, S:** ponderosa *(sic).*

[248] **D** omits "ut melius effluat aqua." **W** omits "melius."

[249] **C:** exibit *aqua tota.* **D:** cum infrigidate fuerit. **S:** si infrigidata fuerit, exibit. **V, Z:** si non infrigidata fuerit, exibit; similarly **P** and **I.**

[Chapter 8]

Quod modo dicturus sum, preeunti similatur ingenio. Nec fit hoc nisi [quia][250] locus vacuus esse non potest. Quia si evacuetur aer, statim succedit aliquod corporum, que ipsi aeri commiscentur, quia [suapte] natura[251] impelluntur. Et hoc quidem asserunt professores sciencie naturalis, quibus similiter opinamur. Unde et manifestabitur, quod[252] locus vacuus esse non potest ab aere vel ab alio corporum.[253]

Cuius exemplum est, ut infundamus aquam in vas, quod sit "a", in cuius medio fiat stans quoddam,[254] quod sit "b", ad modum candelabri super aquam erectum, et in eius summo[255] ponemus candelam, que sit "c", ardentem, super quam convertatur vas, quod sit "d", ita, quod eius os prope aquam sit. Et locus candele in medio vasis sit. Quo facto, paulo post videbis aquam, que est in vase subteriori, ascendere sursum ad vas superius.

Quod quidem non accidit nisi propter hoc quod memoravimus. Quoniam vero aer in illo vase contentus periit propter ignis ascensionem,[256] non potest propter ignem[257] durare. Postquam perierit aer ille, propter motum ignis et quidem elevabit ignis aquam,[258] secundum quantitatem illius, quod peribit de aere. Et hoc accidens simile est illi, quod contingit canali predicto.[259] Scilicet in hoc vase supraposito aer consumitur quia inveterascit, ut ita dicam, propter ignem extenuatus.[260] Et ideo elevatur aqua succedens et subintrans, locum eius implet, qui vacuus fiebat.[261] Et hec est figura huius rei.

---

[250] **S** omits "nisi quia." **N** omits "quia." **C, D, W, Z** have both words. **K** omits the entire discussion and reformulates the description of the experiment. In **P,** "nisi" is added by a corrector.

[251] "Suapte" (sua sponte) is plausibly used by **V.** Inept modifications appear to be used in **N:** sui per naturam; **C:** sua capita; and **S, W, Z:** sua parte.

[252] **M** and **R** omit "que ipsi aeri . . . manifestabitur quod."

[253] **C:** corporum solidorum.

[254] **C:** sit quoddam instrumentum.

[255] **C:** fundo.

[256] **V:** accensionem.

[257] **D:** cum igne. **V:** propter ignem extenuatus.

[258] **D:** continget quod elevabitur aqua. **R:** elevabit ignis aquam. **V:** . . . evacuabitur . . . A confused sequence of words is used in **Z.** The text in **N** reads: qu' elevabe' ig's aq̄.

[259] **M** and **R** omit this sentence.

[260] **D:** extenuatum. **Z:** inveteratus et extenuatus.

[261] **D** omits "subintrans" and "qui vacuus fiebat."

[Chapter 9]

Constat quidem ex premissis quod, posita una parte canalis[262] in vas plenum aqua, si quis suxerit ex alia, detrahetur aqua donec omnino evacuetur cum canali illo. Unde tale proponatur exemplum.

Perforetur vas "a"[263] in lateris fere summitate "b"[264] et imponatur canalis "c"[265] usque prope fundum vasis, firmiterque in foramine ligetur aptatum cum vase. Et sit pars canalis exterior aliquantulum prolixior, inferius descendens quam interior.[266]

Dehinc impleatur vas aqua,[267] quousque elevetur supra summitatem canalis, et incipiet emanare. Quia enim aqua altior est, cum impletur interior pars "c" et additio aque impellit aerem, qui est in "c", et impulsum sequitur, non cessans effluere, donec prorsus evacuetur quicquid aque fuerit in vase. Et hec sit huius rei figura.

[Chapter 10]

Quidam ex huiusmodi artificiorum opificibus conati sunt, ut facilius[268] facerent genus vasis, intus latentem ponentes canalem hoc modo.[269] Sit vas "abg" foratum subtus in medio, ubi sit "g", et intromittatur[270] per foramen canalis ibidem firmiter aptatus, qui sit "gd", directe erectus fere ad summitatem "ab". Huic coaptetur circumpositus canalis alterius add [ad] vas [*firmatus*], qui sit "etk"[271] ita tamen ut relictis ad "t" et "k" foraminibus, aqua possit ibidem introire,[272] si fuerit infusa. Que paulatim ascendens a "kt" versus

---

262 **L** omits "constat . . . quod" and continues "in vase pleno." **S:** posita candela *(sic)*.

263 **C:** perforentur vasa.

264 **C:** in lateribus, summitate pixidis. **D:** summitate "c." **L:** in latere versus summum. **Z:** in lateris fere summitatem quod sit "b".

265 **C:** canalis curvus. **W:** canalis "d" (so also the drawing).

266 **D:** inferior descendens quam interior. **S:** inferius descendens quam inferior. **L** omits the clause.

267 **C:** Deinde impellatur aqua.

268 **D, W:** subtilius.

269 **L** omits "latentem . . . hoc modo." **C:** in cuius concavitatem ponentes. **V:** ponentes daspnaltum daspnaltum *(sic)*. **K** omits Chapter 10.

270 **D, W:** intromittatur aqua *(sic)*.

271 The text of **N** has "formas," apparently a misreading of "firma's." **Z:** firmatus. In the next few lines and chapters, **D** seems to have further difficulty in reading the prototype.

272 **L:** introire a vase in canalem. **D:** si vasi fuerit infusa.

["e"], cum add [ad] "e" pervenerit, per "dg" descendit. [Iustitionarum] hoc est vas.[273]

[Chapter 11]

Post hoc autem aliud est dicendorum elementum. Quod est, ut fiat vas, quod sit "ab",[274] in cuius fundo minuta fiant foramina, ubi est "a".[275] Cum igitur[276] impleveris vas illud per impulsionem per aquam, si obstruxeris os vasis cum digito, nihil emanabit.[277] Sed quando removeris digitum ab ore, incipiet aqua effluere.

Cavendum autem est, ne nimie capacitatis sit vas, propter ponderositatem aque, que descendet, si nimia fuerit.[278]

[Chapter 12]

Sed si vasi huiusmodi [tali],[279] pleno, folium vel aliquid tale ad os posueris et, cum manu id ori applicans, vas totum converteris, manebit ibi folium pendens aliquamdiu, tamquam esset affixum.[280]

Vel si etiam cribrum ori apposueris et[281] converteris, nil penitus de aqua exibit.

Cuius ratio est eo quod, hinc et inde aqua accurrens,[282] minuta illa implebit foramina. Nec amplius descendet, quia non potest aer subintrare,[283] cum aqua non exeat. Et prius ostensum est quod nil vacuum esse potest. Cum igitur

---

[273] **C:** est enim vas iusti[ti]anorum cuius hec [est] descriptio. Similarly **G, H, S, W, Z.** Senseless terms occur in **D:** et hoc est vas iatrici ornatum; **V:** et hoc est vas intracatum. The scribe of **P** exclaims: "et hoc est vas institonatum. Et hec est eius figura, quam prohibiscimus." **N** reads: ". . . instatonarium, et hoc . . ."

[274] In **D** and **W** the words "fit vas" are repeated as a gloss or legend to the drawing. **K:** Si vis facere aquamanualem. **S:** Experimentum de cantaplora. **Y:** Foramina minima.

[275] In the drawing of **N,** no letters are shown. In **C, D, W, Z** the letters are "ch."

[276] In the text of **N,** "cum igitur" is repeated.

[277] **C:** Nil exibit inde aque.

[278] **C:** descendit *vi aut impetu* si nimia fuerit. **D, W, Z:** descendit *vi* si . . . **L:** vi descendit si *nilta* fuerit; the meaning is unclear. **V:** descendit *nisi* nimia fuerit; an evident error.

[279] "tali" is cancelled; so are some letters of "huiusmodi."

[280] **C:** tanquam *illis* esset affixum *foraminibus.*

[281] **C:** aut si etiam cribrum *ipsi orificio* adjunxeris et *vas* converteris. **L** omits the clause.

[282] **C:** discurrens. **D, W:** occurrens. **V:** concurrens.

[283] **D:** subire. **W:** subire, and above it: subintrare.

aer non habet, ubi subintret, manet aqua stans, non recedens. Cuius hec est figura.

[Chapter 13][284]

2.a. Ut restringatur aqua per deorsum, subtus insuffla. Ut autem refluat, suge.

[Chapter 17]

Fiat et vas[285] ore stricto, quod sit "ab", perforatum in fundo, ubi sit "d". Per quod intret ortogonaliter canalis, in summo recurvus, pertingens fere ad collum vasis, qui sit "cde". Et sit pars eius, que est "de", in vase erecta,[286] reliqua vero pars, que est "cd", sit exterius extensa. Cuius extremitas, que sit "c", intret in vas aliud, quod sit "ghz" sitque subtus foratum. Et aptentur ambo hec vasa stantia super quoddam concavum, quod sit ["mn"],[287] ita quod aqua emanans a fundo vasis "ghz" in loco "t" possit effluere,[288] et non alias. Aptetur quoque canalis alius, a vase "ab" exiens in loco "r",[289] extensum [extensus] ita, quod capud eius, quod sit "p", elevetur super vas "gh", qui canalis [sit] "pqr".

Cum igitur impleveris aqua vas "ab", minus altitudine "de", et obstruxeris os eius cooperculo, quod sit "xy", non cessabit aqua fluere per "p" in vas "ghz", donec, impleto super quod constituitur concavo, fiat aque ascensio usque ad "c". Quo aqua repleto, non habebit aer, quo[290] intret in vas "ab" et omnino cessabit aqua effluere per "p", usque evacuetur pars "c". Et tunc iterum[291] exibit ut prius. Et sic semper manebit eiusdem altitudinis.[292]

Chapter 18]

Ad idem quoque fiat aliud vas eiusdem generis, sed levius, quod sit "ab", super duas columpnas "d" et "e". Os sit "c" et canalis exiens "rp". Sitque[293] una columpnarum concava, per quam intret in vas "ab" canalis extensus ad vas "ghz" et subintrans in loco "z", qui sit canalis "qfz". Apteturque vas

---

284 A line, added at the top of the page.

285 **C:** Nunc fiat vas aliud. **K:** Fiant duo vasa. Unum habeat os strictum et clausum cooperculo, aliud quantulum est descopertum . . .

286 **D:** erecto.

287 **D, W, Z:** "m, n." The scribe of **N,** evidently misreading "m," writes: "in n."

288 **C:** emanare. **D, Z:** influere.

289 Text of **D:** "t." Drawing: "r."

290 **C, V,** add: aer. In **L** and **R** the text breaks off near this point.

291 **D:** et tamen.

292 In **E, I, L, V, Y** the chapter normally counted as No. 21 is inserted here: tradition L-II.

293 **C** omits the clauses: "'c' et canalis exiens 'rp.' Sit-".

"ghz"[294] suppositum ori canalis egredientis, ita quod aqua emanans a "p" decidat in ipsum [in plano].[295] Impleto igitur vase "ab",[296] secundum altitudinem "r" et obstructo ore eius, quod est "c", exibit aqua per "p" in vas "gh", quosque impleatur. Et fiet omnino hoc in vase, ut in predicto.

[Chapter 19]

Et ad idem fiant vasa talia "ab" et "ghz" cum suis canalibus "pqr", "cde", omnino ut prius, nisi quod interponatur paries, qui sit "st". Impleto igitur vase "ab" fere usque ad "p" et eius ore obstructo, quod sit "o", descendet aqua per "cd", emanans ad "c" et ibi subintrans in vas "ghz", donec perveniat ad "r", et ita erit ut prius.[297]

[Chapter 20]

Item fiat vas aliud, una superficie contentum, quod sit "abc" et habeat a lateribus duo [canales] exeuntes, que sint "cd", "be". Habeat quoque canalem erectum, descendentem interius in vas "ghz", firmiter in utroque vasorum fixum in locis "l" et "m". Qui sit canalis "klmn". Sintque particule quedam vasis "ghz" ad modum crucibulorum, a lateribus exeuntes, subtense canalibus "cd", "be", queque suo, que sint "gt", "sz". Cum igitur impleveris aqua "abc", minus altitudine "n", effluet liquor[298] per aperturam "cd" [in] "sz" [et] per "be" in "gt", hinc et inde influens in vas "ghz", donec perveniat ad extremum canalis "ck". Quo repleto, cessabit emanatio a "d" et "e".[299]

Sit, inquam,[300] liquor in vase "abc" oleum et aptetur licinium in "ghz", vel papirus, ubi est oleum. Et secundum quod ardendo consumpserit de oleo in "ghz", descendet paulatim ab "abc" per "d" et "e". Hec autem[301] est eiusdem generis et ad idem valent.

---

[294] **C** omits: "Apteturque vas 'ghz'." **W** omits this sentence and the next one up to "r."

[295] **D, S, Z** omit "in plano."

[296] **C** omits: "decidat in ipsum . . . Impleto igitur vase 'ab'." It seems the reading of the prototype was difficult, but the scribe of **C** recognized the sense of the description; see note 297.

[297] **C** adds: "Semper vas illud eiusdem altitudinis et quantitatis aque, prout ante, preservat illud."

[298] **V:** *non* effluerет liquor positus *(sic)*.

[299] **V** omits the sentence "Quo repleto . . .;" **Z** and others have minor alterations of verbiage in the two preceding sentences.

[300] **C, D, S, W:** Sit itaque.

[301] **C, G:** hec omnia. **D:** hec autem quatuor instrumenta. **S:** "et hec 'c'" or "et hec 4." **F** omits this detail.

[Chapter 21]

Et fiat vas subforatum et canalem subterius habens, ut sit "g", quantalibet eius parte 2 vel quatuor disiunctum claustris,[302] que sint "ab, cc, d et e", quorum quodlibet sit et subterius et superius perforatum. Quibus singulis quibuslibet liquorum repletis, si foramina superiora obstruxeris, nichil emanabit per inferiora. Quocunque vero superiori aperto, descendet liquor effusus per subterius, emanans per canalem vasis qui est "g".

Explicit liber Phillonis *de sp[iritu]alibus ingeniis.*

---

[302] **D:** quatuor vel quotlibet. **V:** 9 vel quo[t]libet. **L:** 6 vel quotlibet. **Z:** e vel quotlibet.

Comparative Notes on Western Philo Manuscripts

We must now, at least briefly, consider the western copies of Philo's *Pneumatica*, other than the preceding CLM 534. These copies differ from one another in some of their characteristics, additional to their textual variants, which we have noted in the collated transcript.

I will begin these notes with the Cambridge-Pembroke copy 169, manuscript **C,** which appears as one of the most important parts of the western tradition. It is one of the earliest copies and it has some of the most incisive drawings, to be found in the existing record. A reproduction of these drawings follows as Figure 9.

Cambridge-Pembroke Ms. 169 (**C**)

As mentioned in preceding notes on documentary details, the Cambridge manuscript has been held by some paleographers to belong to the early 1200's—the earliest time from which we have any Philonian copies, now existing. It is part of tradition L–I, and is entitled *De subtilibus.* The introduction to the text, as copied in this manuscript, reflects a dedication to "amtomane:" a name derivable from the original "ariston" only by a series of erroneous copying efforts as to this detail. A somewhat later origin of the copy is suggested by this fact, and also by a note, inserted on top of the first page: "*Ricardus* le sire de Stoketoun et *Walterus* frater ēius, die Veneris post festum Sa. Augustini," without year. Men of these two names flourished in the Oxford and Cambridge regions, 1302 to 1346; Richard was a schoolmaster and one Adam de Stocton was a scribe, c. 1375, according to local, biographic registers.

Contained in this copy are Chapters 1–12, 17–21 and a brief Gothic postscript traceable to Chapters 55–57. The postscript will be transcribed and discussed hereafter. The copy omits the drawing of Chapter 6, shown in two different forms in other copies of the Latin traditions. Other modifications of the basic Philonian tradition are shown by the text of **C,** which has a number of amplifications, omissions, and other changes from the Latin standard text, and from the probable Latin prototype text, as shown in the collated edition, above. Substantially all of the textual modifications, in **C,** are of a clarifying kind.

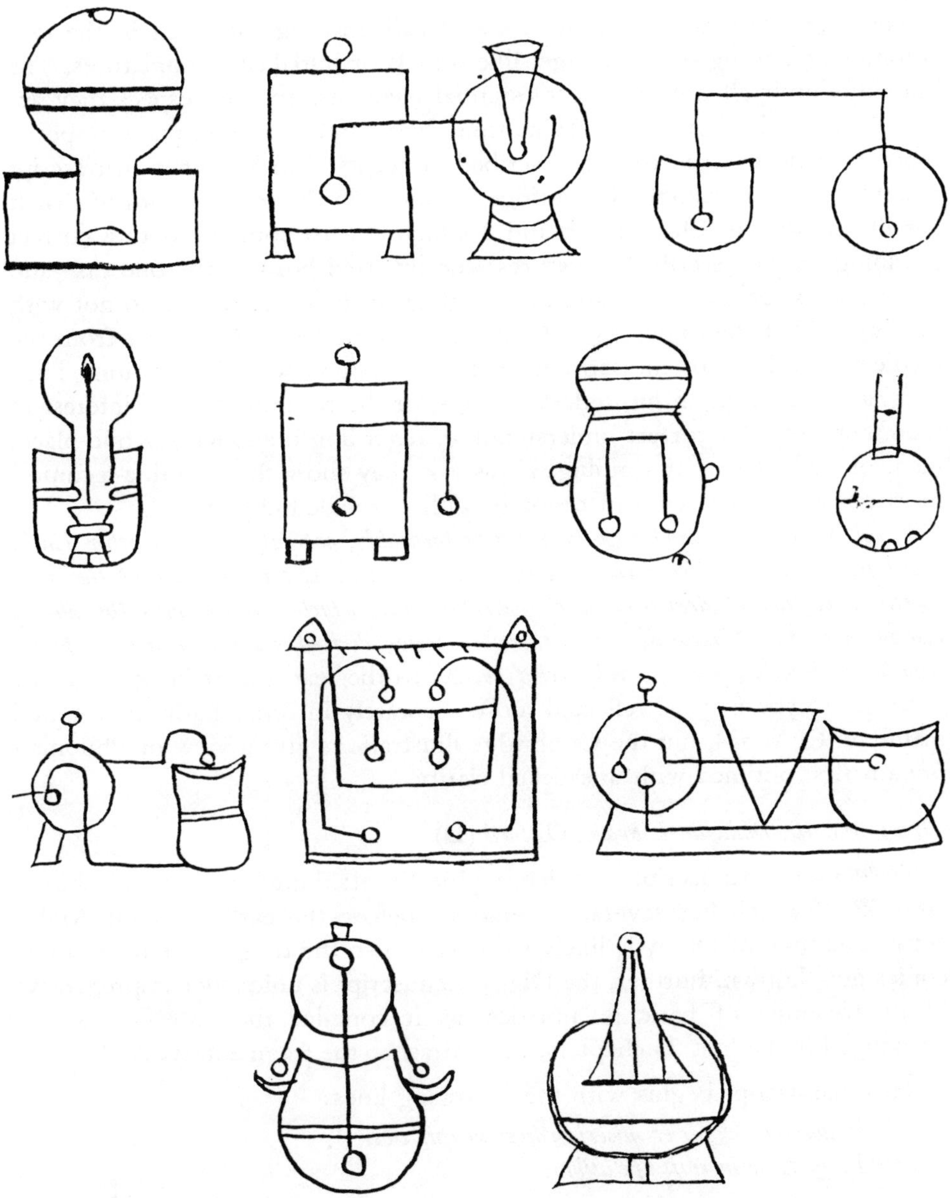

Figure 9. Cambridge-Pembroke Manuscript 169.

First line: Drawings of Chapters 2, 4, 7
Second line: Drawings of Chapters 8, 9, 10, 11
Third line: Drawings of Chapters 17 18, 19,
Fourth line: Drawings of Chapters 20, 21

The illustration of **C** (Figure 9) uses small drawings, inserted in the two columns of writing on each page, like initials or initial-sized miniatures. The drawings omit all but the most essential elements. In this respect they are in diagrammatic form. At the same time their contents are, in many respects, more realistically true to the described concepts than all other manuscript illustrations. For example, Cambridge is unique in the Philo record in that it has illustrations to Chapters 4b and 8 which at least come close to a correct outlining of the described structures: the inverted horn in the one case and the supports for the candle and its container in the other case. I do not wish to suggest that the drawings of Cambridge are perfect. They suffer from the medieval tradition of indiscriminate use of plan view and elevation. They also have the common misunderstandings, or the common lack of interest in representation of a correct understanding, regarding the exact, relative placement of water levels and piping. However, they show the scientific-technical message of the first score of chapters with unrivalled clarity.

In this respect the Cambridge copy differs favorably, not only from the other Latin copies but principally from the Arabic tradition. This is true in spite of the more careful execution of drawings in the Arabic copy, which is done with the aid of instruments. The Cambridge copy therefore seems derived from a Latin prototype translated directly from a Greek archetype. As to the text, the archetype and the Latin prototype may be reflected more accurately in other Latin copies, and mainly in CLM 534, but the Cambridge illustrations give a view of Philonian pneumatics, outlined with unexcelled clarity.

Digby Ms. 40, Bodleian Library, Oxford (**D**)

The Digby manuscript is paleographically attributed to a time, shortly after 1200, which lies several generations before the earliest extant Arabic copy. The time of Digby is likely to precede also the time of the other Latin copies now known. Further, the Digby manuscript is uniquely comprehensive in its coverage of basic pneumatics, as it contains the unbroken set of drawings for the first 23 chapters, in contrast to the fragmentary, Arabic sets.

This manuscript begins with the following lines:

In nomine dei pii et misericordiosi incipit liber
Philonis de ingeniis spe'alibus.

Edward Kente[303]

[Qu]ia tuum, amice mi Arzothont, iam novi desiderium
ad sciendum *ingenia subtilia* . . .

---

303 A signature in Elizabethan script. Identiy unknown. Invocation in Gothic cursive script.

The drawings are irregularly placed. Those of Chapters 6, 7 and 17 extend across the text, as in A. S. 3413. Those of 2, 4, 8 to 13, 15 and 18 are on the lateral margins. Others are placed, without order, in blank spaces left for this purpose. The sequence, at No. 21, and the form, at No. 6, are those of tradition L-I.

Although the Digby manuscript is given a relatively early date, its contents create the impression that its writer and draftsman used an old and poorly preserved prototype. For Chapters 13 to 16, 22 and 23 these men found drawings but no text; apparently they also failed to find the small gloss for Chapter 13 which the writer of CLM 534 still discovered, some decades later, in the prototype material. They found all drawing elements for Chapter 22 but placed them on different sheets, in a way that indicates their failure to recognize the presence of parts of a single chapter. They may have copied these parts from a poor source document, the drawing sheets of which were in shreds. This impression is reinforced by the odd sequence of some chapters in the Digby copy: 13 – 15 – 11 – 14 – 16.

The writer of Digby encountered additional difficulties with the text. He recognized some of the problems, as is shown by the collation (for example, at notes 219, 222 and 258), but he handled other problems with the use of technically erroneous or at least very questionable expressions (notes 227, 244, 261, 270, 271 and 285).

In spite of these defects, wich some of the other manuscripts avoided, the Digby copy was successful. Some of its drawings, totally omitted by other copies of the Latin prototype, reappear in a few of the later Latin copies, including CLM 534 and Palatinus 1377; and one of its characteristic drawing features, which is lacking in all other Latin copies—the showing of "the Sun" in Chapter 7—recurs in a later work of Fludd. The work became important for the development of a basic, modern instrument of observational science: the thermometer.

CLM 444 in Bayerische Staatsbibliothek, Munich (**M**)

The smaller one of the two Philo copies in Munich, CLM 444, is a fragment which only has the texts of Chapters 1/2, 4, 8 and 17. It has the pertinent drawings, and also those of Chapters 9, 18 and 20. These latter drawings are randomly placed in a blank space at the end; the others extend across the text. To this extent the form of the manuscript resembles that of the Digby copy. In view of the lack of the critical Chapters 6 and 21, we cannot decide whether CLM 444 belongs to Latin tradition L-I or L-II. The title may be read: *de sp'alibus ingeniis n[atur]e.*

The contents of CLM 444 resemble the corresponding parts of other western copies, but the drawings show a fairly unique form of ornamentation.

In Chapters 18 and 20, CLM 444 uses stylized animal feet at the feet of portable devices: the feet of a water-receiving basin in one case and the feet of the self-refilling oil lamp in the other case. This well-known element of ancient decoration is otherwise lacking in most of the Philo drawings, both eastern and western.[304]

Sloane Ms. 2030 in the British Museum (**S**)

Here we have another part of tradition L-I, this one probably copied in the XIV. century. Dedication: "marzotom." Title: "Liber Aristotilis *De conductibus aquarum*." Scope: As in CLM 534, except that it omits Chapter 13. Placement of drawings: in side and bottom margins around the text, approximately as in CLM 534. The contents of the text,[305] however, are *different* in many details from the consensus of manuscripts, illustrated by CLM 534, as shown by the present collation. **S** shares some of its many variant readings (or perhaps we may say: most of those which are relatively plausible) with **C** and with the Palatine manuscripts **WXZ,** but not with **N** and the consensus of others. Of the late manuscripts, **K** and **V** have some variants of **S.** Perhaps these facts would justify establishment of a sub-tradition, within the basic tradition L-I.

The drawings of Sloane are executed with greater care than they are in many other copies. Only in **S,** the Palatinus manuscript **Z,** and the Arabic copy **A** does the draftsman consistently use drafting tools (compass and straight-edge); all other drawings are freehand. As to their contents, the drawings of Sloane (published by Rose) differ in some details from the consensus of the Latin tradition, and differ more widely from the Arabic tradition. Nevertheless they share various errors, evidently taken over from the prototype and archetype. For example the design of piping, in Chapter 10, is entirely misleading.[306] This particular error was noted by a later glossator of Sloane, who corrected the defect by adding two improved sketches, at the bottom of the last page.

Add. Ms. 34,113, British Museum, London, The Italian Philo (**I**)

This translation was recorded about 1445, as will be shown in a separate note about its text, hereafter. It renders the ancient dedication *(mi aristō –*

---

[304] However, it appears at the feet of the tank of Chapter 21, in **Y.** In ms. **A,** see Figures 57 and 58. Also see note 472.

[305] Published by Rose and Schmidt; also, with a good technical-historic analysis, by Gerland-Traumüller[79].

[306] As is true also of **C.**

mi arcito – mi argato) by the words "*mio aghate.*" The title is fairly normal: *Ingegni ispirituali.* The contents have the standard scope, also found in the Sloane copy of tradition L-I and in the Vaticanus of tradition L-II. The chapters are arranged in the sequence of tradition L-II. They share typical limitations and confusions of other copies, for example in the illustrations of Chapters 10 and 18 (Figure 10).

In keeping with characteristics of tradition L-II, and mainly with manuscript **P**, Chapter 6 has an illustration, in the Italian copy, which adds materially to the standard contents of tradition L-I. The similarities between **I** and **P** extend to a number of text details, as will be noted in connection with the text of **I**, hereafter. These similarities also extend to the legends on the drawings of Chapter 6, as follows:

Drawing in **I**:	*Drawing in* **P**:
Vaso perforato	Vas foratum
Vaso non perforato	Vas non foratum
Stagnio	Aque profundum / Contigua aquarum
Vaso dove / l'aqua discende /	Vas in quod descendit aqua
dopo il scorre / in alto	postquam ascendit in altum.

The impression arises that **I** is translated from **P**, but that **P**, at the time, was more complete than it is now.

The Italian Philo copy is part of a comprehensive copybook from the circle of Mariano Taccola and Francesco di Giorgio Martini. Along with other material from this circle, it documents the transition from the relative, technical stagnation of the middle ages to the Renaissance of ancient technology in the XV. century. Its medieval character may be seen in its tendency to preserve, substantially without change, the ancient, long antiquated handbook material: a tendency which it shared with the earlier, western and eastern copies of Philo's work. The innovative nature of the document lies in the fact that it, for substantially the first time, talked intelligibly to artisans about the basis of the technology used by them.

Palatinus Latinus 1377, Vatican City (**Z**) *and related Palatine Copies*

The Palatine copy 1377 in the Vatican is as carefully illustrated as Sloane 2030, which it also resembles in other respects. This copy is as comprehensive as Digby 40 except that it omits the drawings as well as the texts of Chapters 13 and 22. Its drawings are executed with greater realism as to details, such as the relative locations of liquid levels and pipe ends, than is found in most of the Philo manuscripts, either Latin or Arabic.

Up to Chapter 21, each drawing is placed substantially at the bottom of its chapter, as in the Arabic tradition. At the end, in about the same way as

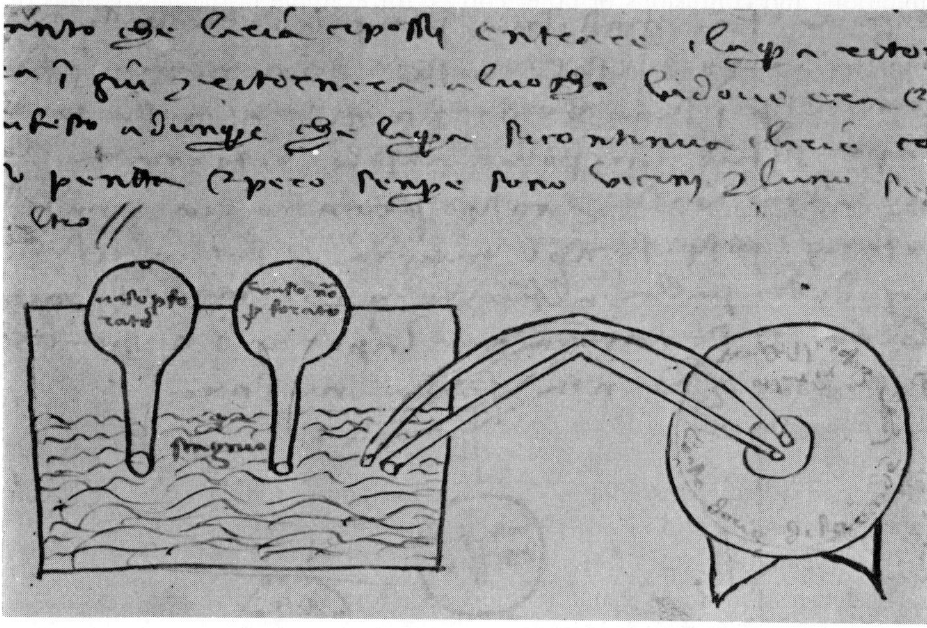

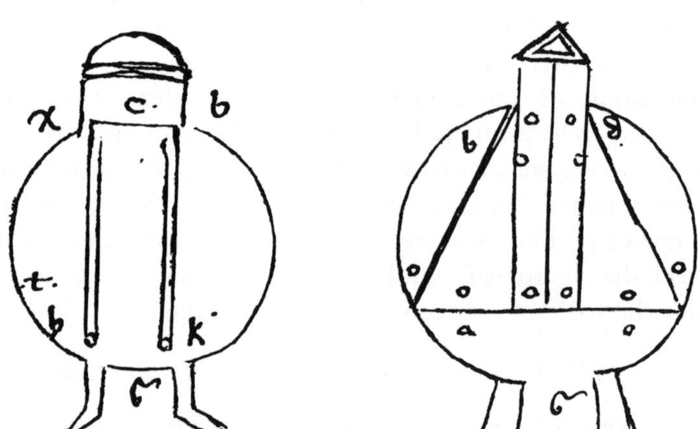

Figure 10. Add. Ms. 34113, British Museum (Italian Philo).
Drawings of Chapters 6, 10 and 21

in Digby 40 and CLM 444, there follows a set of undescribed drawings. Here in Palatinus 1377 they belong to Chapters 14, 15 and 23.

The Initia of this Palatinus are similar only to those of Sloane:

> Incipit liber Aristotilis *de conductibus aquarum*.
> Quia tuum, amice mi arzathonii, iam novi desiderium . . .

The text is followed by the note:

> Explicit Philonius *de aquarum ductibus*.

These Initia appear to be copied, here and in Sloane, from a prototype without invocation, with dedication of the "*miarzatō*" or "*marzatō*" type, with attribution to Aristotle *or* Philo, and with a title of the "water duct" type. Thus they are *copied from a Latin prototype conceivably related to the Arabic tradition but not at all closely related thereto.* With reference to some other copies, this Latin prototype is hypotesized in Chart 7.

This prototype of the Palatine group was not necessarily the prototype of all Latin copies. It may have been the prototype of a subtradition, forming part of Latin tradition L-I, as noted in the discussion of the Sloane manuscript.

In any event the illustrations and descriptions of **Z,** copied from this prototype or sub-prototype, exhibit the basic character of being, conceivably, related to the Arabic tradition and to some of the Latin subtraditions, but not at all being closely related thereto. For example, **Z** shows an addition to Chapter 6, totally independent of the other copies. The illustration of Chapter 11, which is also in **Z,** shows some limited similarity with the Arabic representation of a clepsydra sieve, with a cup below it, but it is very different in the construction of the sieve. In A. S. 3713 the sieve resembles the ancient artifacts with a hollow, stirrup-shaped handle (Figure 2c); the Latin copies, like the Hero manuscripts, follow different, ancient traditions (Figures 2a and b). According to Chapter 17 of the two manuscripts, A. S. 3713 and Palatine 1377, the respective prototypes may have descended from source material which plausibly, and in contrast to other copies, showed the upper one of the wine-receiving vessels as a small cup, manually to be dipped into a lower one; the descriptions of the Arabic and Palatine manuscripts differ.

The impression thus arises that specific features of both the western and the eastern copies are affected by independent changes from a common archetype, or by independent uses of different, Greek archetypes, and are also affected by a great variety of changes and selections in the derivation of the actual copies from the prototypes.

Chart 7. The Palatine Group of Philo Copies

	Common Prototype of **S** and **Z**	Palatinus Latinus 1377 (**Z**)	Palatinus Latinus 1332 (**W**)	Palatinus Latinus 1369 (**X**)	Palatinus Latinus 1375 (**Y**)
Author:	*Aristotle or Philo*	Aristotle or Philo	Philonius	Philonius	Philo
Title:	*De conductibus aquarum*	*De conductibus aquarum*	*De sp'alibus ingeniis*	*De sp'alibus ingeniis*	*De spu'alibus ingeniis*
Dedication:	*marzatoni*	miarzathonii	miarzatoni	marzatō	ioconthī, corrected to marzatom
Invocation:	*None*	None	None	In nomine domini pii et miseri-cordissimi	In nomine [domini] pii et misercordis
Tradition:	*L-I*	L-I	L-I	L-I	L-II
Probably copied:	*XIII. c.?*	*Early XV. c.*	*Between* **Z** *and* **X,** *correcting the title*	1455, after **C, D, N**; *adding the invocation*	1493, after **X,** *correcting the dedication*

Krakow Ms. 568 (**K**)

The Latin Philo texts considered up to here reproduced substantially the exact, technical content of the source documents. Only in areas such as the introductory formulas did the copyists reformulate the contents with some freedom.

Upon the beginning of the Italian Renaissance, copyists in some scriptoria became slightly more independent of the source documents and more ready to restyle their technical descriptions. The Krakow manuscript illustrates this process.

This is a late XV. century version of the Latin Philo. Without invocation or dedication, it covers Chapters 3, 4, 7, 8, 9, 11, 12, 17, 18 and 19 (fo. 104r–v). It then continues with a text on the building of wells and mills (fo. 104v–106r). This text appears to be derived from Vitruvius. It is hardly illustrated. The Philo chapters have the usual number of illustrations; these are rather vague in a technical sense but are graphically original. Their descriptions are restated in modified terms, but without change in substance; literally unchanged terms occur in some lines. An Explicit designates this material as "Liber Philonis hebrei sapientissimi *Super quibusdam mirabilibus ingeniis et apparentiis naturalibus*, de quo dixit Rabanus in prologo super *librum sapientie*." The reference to "apparentiis naturalibus" may have to do with the pneumatic and hydrological phenomena, described in the book; the reference to "Philo Hebraeus" is a plain error.

At the end of the volume (fo. 106v) the manuscript presents an excerpt "ex libro Heronis *de inani et vacuo*, quem inscripsit Landfridus." It contains Chapter I, 4 of Hero's Pneumatica, with vague illustration, and the approximate contents of Chapters I, 20, 21 and 24, without illustration.

The Krakow manuscript thus reflects the transition to a new attitude with respect to Alexandrian pneumatics. As in Gothic times, Philo's Latinized text was preserved in substance, and was supplemented by texts of Vitruvian kind. In addition, the Philo copy of Krakow (copied in Prag) provided a Latin translation of Hero, in parallel with the Hero translations of Burana and Valla, also made in the late 1400's. These earliest, Latinized Hero texts were concentrated on the author's views about "the empty and the void:" views of anti-Aristotelian nature. The minds of physicists became ready for the coming, Torricellian revolution.

Fondo Conventi, Ms. C.S., JX 26, Florence (**F**)

During this transition and preparation of minds, either Aristotelian or Philonian pneumatics continued to be set forth in the established forms, either Scholastic or Alexandrian, by further Latin copies based on traditions L-I and L-II. One of them may be found in the Florentine copy, preserved in the library of a convent suppressed during the Counter-Reformation.

This is a XV. century copy of tradition L-I, beginning:

> Incipit liber Philonis *De specialibus ingeniis.*
> Dixit: Quia tuum, amice Martatom, iam novi desiderium . . .

In general appearance, this manuscript is quite similar to our CLM 534. In scope, it resembles the Palatine manuscripts **W, X** and **Z,** which do not have the fragment of Chapter 13 still found in the Munich copy. The text is very close to the consensus, also reflected by this copy. Like most of the other copies, it tries to reproduce the tradition with accuracy, not with originality.

Geneva Ms. 80 (**G**)

More than a hundred years after the inception of book-printing, and several decades after the first printed edition of an extract from Hero, Philo's pneumatics still were copied by hand. The Geneva copies is one of these late manuscripts. In its text as well as its title *(De subtilibus ingeniis)* and its dedication *(antoniane)*, it appears influenced by Cambridge-Pembroke 169; the text also reflects one of the more conventional subtraditions of Latin tradition L-I. The scope and sequence are as in Sloane.

In this manuscript, the drawings are recorded on two sides of a separate sheet, bearing page numbers 61 and 62 and mentioned on the last text page, fo. 32r, as noted in the documentary introduction. Other late copies, to be described presently, appear to have had similar, separate drawing pages, but to have lost them. In any case the archivists in Geneva, on inquiry, were able to find the drawing sheet in the volume, while the other archives do not seem to have such a sheet. Incidentally the Geneva drawing page, after the Philo illustrations, continues with another matter of pneumatics: a chapter devoted to a medieval version of "Hero's Fountain."

Ashburnham Ms. 210, Florence (**H**)

The Ashburnham text resembles the one of the Sloane copy in calligraphy and in some details, which differ from the consensus. The text also resembles Sloane, rather than the consensus, in that it begins, without invocation:

> Incipit *Tractatus de ductu aquarum* . . .,

whereto it adds: ". . . *plu[vi]arum, dictus ingenia subtilia.*" This copy is unique in that it combines the Latin Philo (scope and sequence of **S**) with the extended copy of an extract from Vitruvius, Book VIII, on water-finding and well-building (bottom of fo. 143v to middle of fo. 144v). Apparently this is the treatise "on collection of rain waters," mentioned in the title. It resembles manuscript **C** in that it also provides the fragment of a Gothic postscript to

Philo; this is written in different Gothic handwriting, on the bottom part of fo. 144v; it is transcribed hereafter. As mentioned, this copy has no (preserved) drawings, but it retains the various textual references to them, and to the reference numbers. Also, on page 143r, opposite Chapter 18, there is a marginal insert (probably written by the second hand): *fi[gur]a n[umero] 12.*

Ashmolean Ms. 1471, Oxford (**R**)

This is another text entitled "*On water ducts*" and dedicated to *marzotō*. It has only the texts of Chapters 1, 2, 3, 4, and 8 (without reference numbers) and 17 (with reference numbers). At the end of Chapter 17 it breaks off with a reference to *ista figura.*

Lyell Ms. 36, Bodleian Library, Oxford (**L**)

This is a fragment belonging to tradition L-II and attributed to the XV. century. It consists of a single sheet, which contains the texts and drawings of Chapters 9, 10, 11, 21, and 18, in this sequence. No Incipit, invocation, dedication, title, or Explicit. The volume containing this sheet, like the volume containing manuscript **M**, also includes the alchimistic work on snake-skin "experiments", copied by Avendar.

Paris, Bibliothèque Nationale, Ms. 7295 (**P**)
and London, British Museum, Egerton Ms. 2622 (**E**)

These are XV. century copies of tradition L-II, which are most conveniently considered together, as they jointly suggest an original chapter-numbering system of the prototype for tradition L-II.

Both manuscripts are mere fragments. Egerton has very abbreviated chapters, no drawings, and no Initia. Furthermore, its pages are badly clipped; they show only parts of Chapters 1–12, 21, and 17–20, in this sequence. The Paris manuscript has Chapters 1–12 and the beginning of the text of Chapter 17. It is extremely similar to the Italian Philo, for example in the drawings and legends of Chapter 6.

The chapter numbering of the Paris and Egerton copies can be shown as follows.

	Paris Copy								Edgerton Copy	
Modern Number	1/2	3/4	5/6	7/8	9	10	11/12	...	17	18
Old Number	1		2	3	4	5	"a"	...	13	14

The Paris copy also has the beginning of the text, although not the drawing, of the chapter now numbered 17; it designates it as number 8. We may conclude that a prototype of tradition L-II had numbers 1 to 6, for the chapters with modern numbers 1 to 12, and numbers 7 to 15 for the modern chapters 13 to 21, in an odd sequence.

Vaticanus Latinus 2975 (**V**) *and Collegio Romano Ms. H.C. 93* (**U**)

We come to two of the latest, and perhaps historically most interesting manuscript copies of tradition L-II. In these copies, **V** and **U,** the scope and arrangement of Philo's work are the same as in the Italian Philo, manuscript **I.** The title is challenging: Incipit *liber Vassor de ingeniis Filonis in ductu aquarum.* Incipit *liber Philonis de ingeniis sp[irit]ualibus.*

As indicated in previous notes, the text of **V** has similarities with **C** and **S; U** was copied from **V;** and this copywork, along with the *De crepusculis* in the same "miscellanea" volume, seems to have come to the attention of Torricelli, perhaps in Rome, thereby perhaps planting the seed of his later discovery.

As also mentioned, these documents are likely to have come to the attention of Kircher, who was Torricelli's contemporary and opponent. In the Collegio Romano, where manuscript **V** was copied as **U,** Kircher was prominent and ubiquitous. Moreover we have the word of his younger associate, who found a book with the unique title of **V,** among Kircher's books. In a bibliographic note, this associate listed "*Vassor arabs* ... [quem] reperi inter libris P. Athanasii Kircheri:" Caspar Schott, *Mechanica hidraulico-pneumatica*, Würzburg, 1657, p. 13. "*Vassor arabs*" may refer to AlKindus of Bassora, confused with AlKindus of Egypt. *This would indicate that the Latin Philo was, rightly or wrongly, considered a translation from AlKindus' book.*

The Greek Fragments

Philo's *Belopoiica* still contains the author's description, in Greek, of a pneumatic model demonstration, witnessed by him as it was given by his teacher.[307] A full and literal translation is as follows:

He (Ctesibius) constructed vessels, similar in form to medicine capsules but without lids. To make them resilient and strong he hammered the bronze of which he made them. First he had made a wax form, and cast them with proper [wall] thickness. He turned the inside, by tool engine, and worked the surface to make it even, straight along the ruler, and smooth. He then correspondingly introduced the bronze piston, which could run through [the so prepared cylinder] and could contact it with a periphery that was made similarly straight and smooth, so that fluid could not be pressed through, even when all [available] force was applied. Do not wonder or doubt that it can be so constructed. For in the flute played by hand, which we call *hydraulis*, the bellows for sending air into the submerged wind chest also is of bronze, and is worked similarly to the aforesaid vessels. Ctesibius demonstrated it for us. He also showed the nature of air and what strong and rapid motion it has. At the same time [he showed] the construction of vessels confining the air, having applied builder's glue around the vessel, having set an end piece on the small disk, and driving in the [resulting] piston, with die and hammer ... When force had thus been applied, and the die then was removed, the piston sprang out of the vessel with great force. Often fire came out with it ... Having constructed the vessels as described, he adapted the form of the catapult to them ... They caused the arms to rock apart with great resilience and to throw the stone quite a remarkable distance.

The authentic illustrations are lost, but the text evidently is clear enough, even if it has suffered abbreviation, to show the scene and its background. It indicates the teacher-pupil relationship: "Ctesibius demonstrated it for us." It shows the novelty and boldness of the operation; Ariston was very apt to "wonder or doubt that it can be so constructed." A scientific-technical

[307] *Belopoiica*, pp. 77f. For the second fragment, translated hereafter: Hero, pp. 412–415. For the Greek reference letters throughout the *Pneumatica*: n. 173f of the preceding, documentary introduction.

experiment of the first rank is here reported, and this is done with uncommon perceptiveness and power.

In the second Philonian statement, preserved in Greek and pertinent to the contents—at least to the secondary contents—of this volume, the reported action was Philo's and the report is by *his* pupil. Hero cites Philo's *Automatopoiica*, in the later book of the same title, about a theatrical performance of the "Legend of Nauplius:"

When the scene first opened, there appeared twelve effigies in three rows. They were shown as Greeks in act of mending the ships preparing to launch them [of the way home from Troy]. These figures were mobile. Some were sawing, or dressing woodwork with axes, or hammering, or working with large or small drills. They made much noise, as might be made in actual truth.

After a proper time the doors closed, and then reopened. Then one saw nothing on the scene but painted air and sea. Before long the ships came sailing by, one after the other. As one disappeared, others appeared. Occasionally dolphins were swimming next to them, alternately diving and surfacing as they actually do. Gradually the sea became stormy and the ships were sailing close to one another.

When the scene was closed and re-opened, the sail-ships were not to be seen. One saw Nauplius [the "Wrecker of Ships"] with raised torch, and next to him Athena. Fire then was made above the scene, as if the flame of the torch was shining above.

The scene closed and re-opened. One saw the ship-wreck, and Ajax [Minor] swimming. Athena was lifted by a hovering machine; this was working from above the scene. Thunder resounded. Lightning hit Ajax, who was in the middle of the scene, and his figure disappeared.

In Hero's report, this text is written in the past tense; in Philo's original book it probably was written in the present tense but otherwise in similar terms. As implied by Hero, this description of the action performed on the scene was followed by the detailed explanation of the instruments and devices. This method of functional description, followed by structural explanation, may be found throughout Philo's books. He fully describes the machines by which "the figures are mobile" with their tools, "the ships come sailing," and "dolphins are diving." It would be hard to find an earlier, systematic use of this method. The method is not used in the writings of Plato, Aristotle, or Archimedes. Considering that the *Elements* of Euclid were new and famous, in Philo's time, it seems tempting to conclude that Philo adapted the Euclidian method of "proposition and construction" to the field of physics and technology.

The Gothic Fragments

In addition to the various Latin copies, which reflect the first score of chapters of the *Pneumatica*, or parts of them, we find a few Latin texts which we may interpret as fragments of the chapters now identified by numbers 55 to 57 of the Arabic version. These fragments have been found, unexpectedly, during the studies preparatory to this edition. They occur mainly in Gothic notebooks devoted to technology. It seems remarkable that the fragments, found in widely dispersed sources, sum up to a restatement of three consecutive chapters, 55, 56, and 57.

In their Arabic versions, and in contrast to almost all other parts of the Pneumatica—the only exceptions are Chapters 7 and 8—, the three chapters have in common that they relate to apparatus for "fire," which was also called "Hermetic" apparatus. *In Chapter 55*, the Arabic drawing attempts to show a box-like float, closely surrounded by a box-like housing, with a tall, central chimney on top of the housing. The specification refers to a lighted candle or "torch" standing on the float, the light of which remains visible though open windows or "doors" in the wall of the housing, while the lamp can rise and fall with the float. *In Chapter 56*, the draftsman shows another box, with hexagonal walls; according to the text the walls could have other forms. They again have open windows, this time for access to the inside of the box. An attempt is made to describe and illustrate an inner, universal suspension (now called a Cardanic joint) either for a censer—a scent-spreading burner—or for an inkpot; the pot or burner is said to be mounted on pivots carried by a system of pivoted brass-rings, so that the pot remains in upright position regardless of different positions of the box. *In Chapter 57* there is further reference to a censer. The drawing shows it as a fire grate, on a boiler kettle which stands on ornamental feet. A steam pipe, with ornamented outlet, goes from the boiler to the fire grate.

Now about the pertinent, western fragments and their historic background.

A postscript to the initial chapters of Philo's *Pneumatica* occurs in the Cambridge and Florentine (Ashburnham) manuscripts. In **C,** on the last folio—directly after the text of Chapter 21 and around its illustration—, we read the following sentence, in the same handwriting as the preceding six columns of text:

Notandum est quod per presens opusculum, quilibet subtilis et capacis ingenii per se ipsum plurima invenire poterit ingenia subtilia et diversa, vel de ymagine ab ore flammam emittente(m), calefacere balneum candela mediante, fervere ollam in loco frigido, calefacere stupam sine igne, accendere candelam nullo vidente, disponere candelam inconsumptibilem.

These suggestions of fire-spitting heads (aeolipiles), candle-heated baths, and fireless cooking vessels and stoves appear related to Philo's Chapter 57 and its ancient-medieval developments, whereas the invisibly lighted and indestructible candle unit may be connected with similar developments of Chapter 55.

In ms. **H,** the postscript is generally similar, but ends with more specific proposals:

[N]ota quod per predicta subtil[is] investigator plurima per se poterit ingenia subtilia reperire. Sicut artificia[li]ter componere balnea subterranea et alia balnea et candelas inconsumptibiles. Et calefacere stufas, atque per italiam, tempore yemali calorem et odorem optimum reddentia, atque ymagines cantates et organizantes. Exempli gracia habes [lucernam][308] plumbeam vel ligneam bene picatam, oblongam, latam, non profundam, oleo plenam, in muro locatam ne videatur, foramen habentem tendentem ad candelam inferius acavatam, que cum semel accensa fuerit, ardebit omne quousquam ab oleo recipit nutrimentum.

Ad hoc ut sola oblata carbonum vel lignorum per totam diem habeas aquam calidam, impleas cufam aqua. Cura fundum foramen habere, in quo locatur anellus[?] eneus cum pilla aliquantulum altius[?] sibi, qua fiat ignis[?] ad placitum, donec aqua in [ejum?] bul[l]ire incipiat.

Evidently some of these terms are copied from a Latin source common to mss. **C** and **H.** They refer to traditions which already were old at the times when these manuscripts were written. The invention of the bath in "subterranean" form is attributed, both anciently and now, to C. Sergius Orata, who flourished shortly after 91 B.C. and thus belonged to a time one or two generations before Vitruvius.[309] He had begun with subterranean heating of fish-ponds, establishing the entire pond on a false floor supported by a system of low pillars. Warm air from the waterheating furnace passed through the underground flue or *hypocaustum*, heating the floor. The identical system was soon applied to the floors of baths and of apartments, and later also to house

[308] Between "habes" and "plumbeam" the word "ymaginem" was originally written, and then erased.

[309] Münzer, in Pauly-Wissowa, s.v. Sergius-33; Saglio, in Daremberg-Saglio, *s.v.* balneum; Thédenat, *ibid.*, *s.v.* hypocaustum.

walls or to a stove: *calefacere stupam sine igne*, **C.** The invention was most successful in the frigid north.[310] From there it was reintroduced into its original homeland; atque per italiam, **H.**

A historic connection between the original content of a Philonian Chapter 57 and the *hypocaustum* of Sergius Orata is suggested by the work of Hero. In this work, prior to Orata, floor-heating or wall-heating air ducts are not shown, but Philo's furnace and water-boiling unit, with steam duct from the boiler to the furnace grate, is developed to some extent by adding the air duct for supplementary heating of the water basin.[311] The abbreviation of Philo, now reflected by ms. **A,** retains only the steam duct for fanning the fire, which was also known to Vitruvius.[312]

A historic connection may also be found between the "*balneum subterraneum*" and the "*candela inconsumptibilis*" of both manuscripts, **C** and **H,** and other details of **H.** At least it seems conceivable that this lamp may be traced to the floating torch unit of Chapter 55, which has the long, wide, and shallow, lamp-supporting float to prevent the lamp from being extinguished by water: *lucernam . . . oblongam, latam, non profundam.*

Here, then, the Arabic manuscript seems derived from an abbreviated version of the original text, whereas the Latin-Gothic tradition retains at least fragmentary details. These details also include the suggestion of a device for projecting animated shadow-images: *ymagines cantantes et organizantes*, from a hidden light source and focussing apparatus: *in muro locatam ne videatur, foramen habentem tendentem ad candelam* . . . The relatively late manuscript **H** also mentions a bath-filling device equipped with a steam-actuated ball check valve: *anellus aeneus cum pilla.* It seems possible that further pertinent details would be found in the work of Omar AlKindus, reflected by Giovanni Fontana if the book of *Automata and Wondrous Machines* could be located.[313]

Because of the fragmentary character of ancient and medieval information on technical developments, the interpretation of the postscripts to the Latin manuscripts **C** and **H** is difficult. Perhaps it becomes somewhat clearer when considered in connection with the further fragments that we may find in notebooks of Gothic engineers. We find further explicit references to Philo in the notebook of Kyeser, a physician-engineer-adventurer who died in 1405.

---

310 Thédenat[309], p. 349.

311 Hero, II, 34; also see II, 6,11,35 and Drachmann, pp. 125–132.

312 Vitruvius, I, vi. 2.

313 Fontana[69], Munich, cod. icon. 242, fo. 20r: "nocturnus ignis" with "ceri . . . qui sucesive accenduntur;" 42v: "fornaces archimistice . . .; stufam pro hieme utilimam;" 43v: "fons virginum . . . in camera clauxa;" 54r: "Fornax;" 57v: "volubilis rota [quae] vertitur a fumo."

His book shows one of the devices which **C** calls "*ymaginem ab ore flammam emittentem;*" Kyeser calls it a "Philoneus."[314] A metallic statuette is filled with aromatic fuel. According to **H,** when the statuette is ignited, it distributes "*calorem et odorem optimum.*" The construction seems related to the figure in Chapter 57, which emits the substance now known as "steam" from its mouth, to fan the fire which boils water to produce the steam: a well-known, ancient and medieval artifact, sometimes called *sufflator.*[315] In addition, the Gothic master remembers a water boiler or heater, independently of this Philoneus; he speaks of the "bath stoves described by the great Philo" and of the "bath construction thaught by Philo's genius." He illustrates the stove and bath construction as a late, northern version of a Roman bath heater, shown with warm water duct and in combination with various furnaces, lamps, chimneys, vapor baths, flasks and physician's paraphernalia.[316]

In earlier, Gothic times, the notebook of Villard, datable about 1250, had shown the brazier of a hand warmer, suspended on the universal joint, described in Chapter 56. This hand warmer or "escaufaile" (chauffe-aile?) had the form of "a brass apple with two fitting halves, containing brass rings and, in their middle, a little brazier so pivoted on pins of these rings that the coals will never drop out."[317] The device may have had western precedents in late antiquity.[318]

This existing set of Gothic fragments shows a dominant concern with burners. Various kinds of candles, censers, miniature furnaces and full-size furnaces are incorporated in the Philoneus figures, the Philo bath heaters, and the lamps and *camera obscura* suggestions; these are described by the Gothic writers and are suggested in the postscripts of manuscript texts.

In the eastern tradition, these fragments reappear as a coherent set of chapters, numbered 55 to 57. They appear less dominated by a single, Hermetic concern: the brazier of Chapter 56 is mentioned only in second place, and its suspension is primarily used for an ink-pot unit, supportable in various

---

[314] Kyeser[3], fo. 95b.

[315] Vitruvius, I, vi, 2; Ewbank[6], pp. 393–407, 569–600; L. White, *Medieval Technology and Social Change*, Oxford, 1962, pp. 90–92.

[316] Kyeser[3], fo. 112b3: "balnea fornelli a magno Philone descriptis;" 114v: "Philonis ingenium balneum sic docet preparari." Also variants, such as those of Johannes Hartlieb, Vienna ms. 3062, fo. 130v–137v.

[317] Chapter XVII of Villard de Honnecourt[73]. The chapter or note continues with a Heronian Fountain, called „cantepleure." All this is shown on a sheet which had shown figure sketches of dice-players, etc.

[318] See, for example, the reference to the *Mappae clavicula* in Sarton, II, p. 349. Earlier origins are possible but have not been documented.

positions without shedding ink. Yet there is evidence, in the east *and* west, of the survival of a group of chapters on burners, rightly or wrongly attributed to Philo.[319]

These chapters then lead to further developments. For example the "fire-spitting figure" of a water heater, mentioned in western manuscript **H** and illustrated in Chapter 57 of eastern manuscript **A** as the snake-head of a "curved pipe," also appears in a notebook of Taccola, as the drawing of the "Brass head of a Tartar."[320]

As in other instances of "Magic and Experimental Science," little or no development of significance for any progress of science or technology can be shown to have followed directly from these traditions. However, it is naive to look for direct progress from minor, surviving fragments to matters of modern significance. For present purposes it seems sufficient to note that various burner constructions were remembered, and were commonly attributed to Philo, in the western and eastern middle ages. In the east they were also attributed to his book on pneumatics.

---

[319] Drachmann, p. 67, who mentions the reference to censers, seems unaware of the Gothic variant.

[320] Taccola, in CLM 197, fo. 69r (Bayerische Staatsbibliothek, Munich): "Nota tartarum aeneum et vacuum non habentem."

The Italian Philo

An Italian translation of the Latin Philo fills the first few sheets of a large manuscript book, preserved in the British Museum at London as Additional Manuscript 34,113.[321] Substantially the entire book, which contains over 500 pages, is written and illustrated in a single, characteristic hand, apparently of the XV. century. It is unsigned and undated. We will call it the work of the Italian Anonymus. According to a title on the old and possibly authentic binding of the tome, the contents deal with hydraulic, military and other machines: "*Macchine idrauliche, di guerra, etc.*"

The Italian Philo will be shown here by a transcript. Some notes will be added, pointing out the ascertainable circumstances of the translation, and to what extent the Anonymus understood the Latin text, or possibly at some points improved it.

The codex was evidently intended to provide information, in the vernacular, about technical matters. Illustrated descriptions of machines and methods occupy most of the first two-thirds of the volume, while undescribed machine illustrations occupy the last third. The writing is in a Tuscan dialect, with Latinizing phraseology. Different parts of the book seem to be done at different times, as indicated by changes of the writer-draftsman's writing and drawing styles, and by his leaving blank pages. The Philo translation is followed by several blank sheets, and then by a tract which deals with water-finding, well-digging and pipe-construction: a medieval extract from the VIII. book of Vitruvius.[322] This extract of the Anonymus, in London, is substantially identical with an extract in the Ashburnham manuscript (**H**) in Florence, which is also preceded by Philo's Pneumatica. There follows, in the London codex, a compilation about chemicals, notes about the prospecting for metals, and a book about incendiaries, and after these, an Italian translation of the long treatise which Mariano Taccola of Siena had completed in 1433.[323] This treatise, like Philo's work, dealt largely although not

[321] "Il libro di Filone *degli ingegni ispirituali*," fo. 1r–8r, London, British Museum, Additional Manuscript 34113, cited here as **I.**

[322] **I,** fo. 9r–11v, derived from Vitruvius, VIII, i to iii.

[323] **I,** fo. 12r–14v, *De la miniera*; 15r–17v, A fare fuocho . . ., A fare polvare di bonbarda . . ., A fare pessimo veleno . . .; 18v–20v, *Liber ignium* (also see Martini[3], pp. 247–250); 22–153 (see Prager-Scaglia[67], pp. 13–17, 27–32).

exclusively with hydraulic and military machines. These evidently were the interests to which the Anonymus also devoted his principal effort.

Taccola's name does not appear in the anonymus manuscript, but reference is made to "Maestro Francesco da Siena," no doubt meaning Francesco di Giorgio Martini.[324] Many of the contents of the anonymous book also occur in Martini's Treatise, and can be read most conveniently in it, as printed editions are available. This applies mainly to the water-finding extract, the chapters on mining for metals, and the numerous chapters from Taccola which the Anonymus translates exactly: Martini's version is in close parallel with them. In detail, it has more or less significant changes.[325]

The Anonymus always gives plain, straightforward translations of the source documents, with substantially identical copies of their drawings, except when he clarifies details of perspective or the like. By contrast, Martini often shows a readiness to modernize the source document; only at times he acts as a copyist or translator, and he is quite scholarly as such. Repeatedly and comprehensively, he works on translations of Vitruvius' ancient work; he achieves distinct progress in understanding and solving the technical problems therein.[326] In various forms, he translates this work.[327] Similarly, his treatise presents an Italian translation and technical modernization of Taccola's medieval Latin and bizarre drawings. He produces his own reformulation and recasting of illustrations, for example in his version of Taccola's work on the construction of mills.[328]

Of course, errors of detail occur in each copy or version or translation. An example may be found in the drawing of Philo's Chapter 7, as rendered by the Anonymus: it designs the *open* vessel as the heated "pila" or *palla*, thereby making the proposed demonstration impossible. Martini's work is less affected by technical errors of this kind.

Martini, like the Anonymus, gives the honor of prominent position to Philo.[329] However, Martini's work passes through a number of phases, and in some of them, probably the later ones, he concentrates more on Vitruvius. We may designate the Anonymus and Martini as a pair of closely allied con-

---

[324] **I,** fo. 129r, 194v (also see Maltese in his ed. of Martini^3, pp. XVI–XXIII).

[325] Martini^3, pp. 161–169 on water-finding; pp. 171–177 on metals; pp. 178–246 on various civic and military structures.

[326] A critical edition by G. Scaglia is under preparation.

[327] For the form of his restatement of Vitruvian architecture: Martini^3, pp. 3–140.

[328] Martini^3, pp. 149–159.

[329] Martini^3, p. 169: "Alchindus, Hermes, Philon." Hero is not cited.

tributors to the renewal of ancient science, and particularly of Philonian pneumatics, in the Italian Renaissance, directly before Leonardo.

Both the Anonymus and Martini always write Italian, although both seem to be fluent in Latin. The technical style of Martini has two characteristics, which make his work fairly recognizable even where his unmistakeable handwriting is not in evidence (he uses a number of scribes). These characteristics are, his Sienese dialect and his Latinized, sententious grammar.[330] Both characteristics of Martini's diction are in evidence also in the Anonymus' work, particularly in his translation of Philo's Pneumatica and of the appendix to it, the water-finding tract.

Martini may have learned his sententious writing from the Anonymus; this writer appears as the older one of the two contemporaries. When he refers to "master" Francesco, no doubt meaning Francesco di Giorgio (on sheets which he probably produced about 1475, a short time after Francesco had become a master painter) he does not imply that Francesco was his, the Anonymus' master. The sheets containing his translation of Philo, and of the water-finding tract, look considerably older than the sheets of about 1475, which contain his Taccola translation. These Philo sheets also have their own water-mark, of the time about 1445, and of the region of Siena.[331] Thus it seems probable that the Anonymus initially wrote the sheets of the Italian Philo and of the waterfinding tract, translating them from a Latin manuscript of tradition L-II which also had this tract, and that he did this about 1445, when Taccola was still alive; Martini then was only six years old. Taccola's friend, Filippo Brunelleschi, also was still alive at that time. The Italian translation of Philo's Chapter 5 may have led him to an utterance of very similar character and topic, which is reported by Taccola.[332]

The work of the Anonymus had excellent success in the XVI. and XVII. centuries. It was then owned by well-known Tuscan builders, who recorded and signed some of their notes in it. In the middle of the XIX. century, the book still was famous in Florence.[333]

---

330 Martini[3], pp. 161–246, *passim.*

331 C.M. Briquet, *Les filigranes*, Geneva, 1907, nos. 6647–6659 ("tulip"). Everywhere else, **I** uses a "ladder" watermark.

332 F.D. Prager and G. Scaglia, *Brunelleschi*, Cambridge, Massachusetts, 1970, pp. 125–129. The date of Brunelleschi's speech, p. 128, may have to be reconsidered.

333 The book was analyzed and praised by A. Angelucci, *Documenti inediti par la storia delle armi da fuoco italiane*, Turin, 1869, p. 84n. On fo. 255v, 257v and 259r, the book of the Anonymus contains entries under the names of Lorenzo della Golpaia, Bernardo Bonaiuti, and Matteo Nigetti.

The works of Martini had similar and greater success. When his ultimate translation of Vitruvius was hardly finished, its new, technical understanding reappeared in the first printed edition.[334] Leonardo then studied Martini's treatise. Subsequent copyists, in the XVI. and XVII. centuries, complimented Martini by massively copying from his treatise.

As we have noted, ideas of Philo also filtered into ideas of Kircher, Fludd, Galileo and others. Galileo reported that he had learned pertinent facts from well-builders, who told him about the limited power of a suction pump to sustain a column of water. It is tempting to speculate that such well-builders knew about liquid "cohesion" and about well-building practices from the Italian treatise of the Anonymus, through the Florentine builders who then owned this treatise. In any event, traditions on physics were becoming possible again, among workers as well as scholars, by this treatise.

NOTE ABOUT THE ITALIAN TRANSCRIPT

The words of the Anonymus are transcribed in phonetically literal form: We write down what we hear when the old Tuscan text is read out loud. We resolve all abbreviations and introduce commas and accents, which are unknown to the Anonymus. We modernize the text as to punctuation and as to the use of accents. As to the use of reference letters: we change .a. and /A/ to "a".

The edges of sheets in the manuscript are worn, so that letters and words are missing. The context usually allows reconstruction of the lost letters or words. These are shown by cursive printing.

The English translation of the Arabic text, provided in this edition, is generally applicable to the Italian text, as to the Latin one. Substantial variants of the Anonymus are mentioned in footnotes.

TRANSCRIPT

[Chapter 1]

(Fo. 1r) Al nome di dio pio e misericordioso, qui comincia il libro di Filone degli ingegni ispirituali. Disse:

Perchè tu, amico mio Aghate, già hai conosciuto e' desiderato di sapere gl'ingegni sottil*i*, alla tua volontà e p[r]eghiere rispondo, alla tua domanda questo libro componendo, sicchè per la tua diligenzia e stu*dio* intenderai il trattato e in ogni cosa che hai dom*an*dato, [d]e' predetti ingegni no[n] ti sia

---

[334] Vitruvius, ed. Johannes Sulpicius, Rome, 1486. Details may be explained in a future study by G. Scaglia.

incommodo l'esem*pro*. Di queste cose così cominciaremo a trattare. *In primis* [dico] che all'umane cose a niuno savio sono nature incognite.[335]

Unde e' filosafi, che nelle cose considerarono naturali, una grande inerente [o]pin*ione* non conseguenti, dissero [che uno] vaso vacuo non sia, siccome essi hanno considerato. O d'aria o d'alcuno altro corpo *è* pieno. Io vero nella presente mat*eria*, disiderando di servire in brevità, nè le loro paro*le* raccontando nè le ragioni contra esse, al presente voglio retrattare, acciò che 'l mio sermone non ent[r]erà in molta longhezza. È manifesto e liquid[o] e dichiarasi, immo manifeste appare, che l'aria è uno de' primi elimenti delle cose, a noi prima manifeste. E questo appare manifesto, non sicondo in nome tanto. Delle quali sufficentemente è detto a me a posare quello che intendo, e confer*me*rò che l'aria è corpo.

[Chapter 2]

Se io piglerò uno va*so*, il quale crederemo ch'egli sia votio, e sia bucato e in mezzo sia ampro e in sommo stretto, come son*o* le guastarde che in Egitto si fanno, e darògli impressione nell'acqua profonda, cioè col buco al *culo*. Nientedimeno il acqua intrarà [sola]men*te quando n'e*scirà parte d'aria. E dopo il uscito *del aria è* l'entramento dell'acqua. Queste di*mostrazioni* saranno in questo, esemp*ri grazia.*

(Fo. 1v) Pigli uno vaso, quale di sopra t'insegnerai, cioè ampro la bocca,[336] nel quale el fondo sia uno forame minuto e turato con cera. E da poi rivolta il vaso a bocca rivercia e porgalo nello profondo dell'acqua. E nientedimeno è da curare ch'egli sia posto rit*t*o. Da nulla parte incrinato sia. E così affondarlo colle mani per insin a tanto che tutto sia sommerso nell'acqua. E così a poco a poco e soavemente l'accann . . .[337] Trovera[i] lo dentro asciutto e in niuna parte, sino in nella bocca, di fuore, bagnato.

Or ecco che ène manifesto che l'aria è corpo. Se egli non fusse corpo, e votio fusse il luogo dentro, ci corrirebbe dentr*o* l'acqua senza niuna proibizione.

E acciò che meglio si dimostri, il vaso predetto sommergete come prima nell'acqua e la cera posta sopra a quello forame leva.[338] Così escirà il aria per lo forame e questo si vederà sensibilemente. Se l'acqua sarà sopra il forame del

---

[335] "Imprimis . . . naturae incognitae," also "immo" in the next paragraph: the translator imitates the stilted wording of the Latin introductory phrases.

[336] This contradicts the preceding statement: "e in sommo stretto." The same self-contradiction occurs in **P,** not in Latin tradition L-I; see note 166.

[337] **P:** ex"ha'. It seems the translator found the abbreviation unintelligible."

[338] A formulation of the type also used in Martini's *Trattati*[3].

vaso, farà le lanpolle.[339] Ed empirà il vaso d'acqua per l'uscire dell'aria per lo forame detto.

E quello che fa l'aria uscire ène il moto e impitito [impeto] dell'acqua, pegniente l'aria in nell'intromissi*o*ne del vaso,[340] che l'uno pegne l'altro. E questa dimostrazione ène che l'aria sia corpo.

[Chapter 3]

È manifesto. Dichiarerò nientedimeno degli elimenti che sono utili in questa iscienza.[341] Alcuni de'savi dicevano il aria di minuti corpi e minute particule, le quali per la sua parvità per lo sentimento del vedere e per alcuno degli altri sensi non si comprendano, *(fo. 2r)* le quali particule no[n] facilemente si segregano. Quando [el]le sono congiunte insieme, non è così. Disse ancora alcuno de' savi che nell'aria ène [la] loro nat*ura*.[342] E messasi il corpo dell'arie connesso, cioè colle sue par*ti* minute e sottili, siccome ancora nelle parti d'ogni co*sa* molle e così ancora d'ogni cosa.

E questo dichiara*vi* già altre volte suffizientemente, dove io feci menzione degli arbitri mirabili. E a questo ci referiamo.

La sustanzia dello elemento umido si dice continuarsi all'aria più che altra mistione di natura tenace, e non è infra loro distanzia. E però avviene molte volte che l'acqua va sopra coll'arie e [fuori] delle vasi, se [benchè] veramente la natura della sua ponderosità vincesi sopra qu*ello*. La quale con poca la pegnie a discendere di sotto, con tutti e'corpi ponderosi. [Q]esta è la proprietà e cons*ue*tudine, che di natura loro discendino in giù.

[Chapter 4]

[a]

Che il acqua molte volte di sopra se levi, è manife*sto*. E attenesi coll'aria elevata, ch'è connessa, sicon*do* come ène manifesto del vaso con che si gus*ta* vino. [Del] quale el capo quando alcuno il terrà in bo*cca*, soffic[a]ndovi dentro, il aria, che v'è, attrae, e quell*o* che ène il corpo molle e liquidissimo, ch'è sotto,

---

339 Between the lines: legato sopra l'acqua.

340 Literal wording in the manuscript: . . . in nel vaso intromisione del vaso. The first "vaso" should have been struck out.

341 **I** amplifies and clarifies the text of **P** and **E** (which omitted the two words before "nihilominus" and thereby made that term unintelligible).

342 The same erroneous and meaningless term occurs in **P:** q' in ae' sit hor' na'. Evidently the scribe of **P** misread īae, (which stood for the correct term "inane"), as "in a'e" and tried to correct it into "in ae'", "in aere," which the Italian translator accepts.

[anche attrae], perchè all'arie si continua, siccome colla pania si *collega* o con altro ligamento.

[b]

Ed anco così si manifesterà.[343] È opportuno che tu pigli uno corno di bue, *de*ntro bene concavato, per insino che sia lar*go* e sottile. Sia di mediocre altezza e la sua forma come una forma di pina rotunda e pinetto . . . Sotto alla parte, la quale si somiglia la fig*ura* pineale, accostisi uno vaso di legno bene *asciutto*, *(fo. 2v)* acciò che si facci bene quello che noi vogliamo. E così bene s'adatti che in nulla parte il aria possi ispirare. E così sarà la quantità del corno, a modo e similitudine del vaso.

Da poi convertesi questo corno, istante in quello vaso, che la bocca sia ampra come uno bichiere.

Item ponerette uno altro vaso, di piombo, sotto a questo,[344] bene attato acciò che quando sarà di bisogno, niente d'inde esca d'arie. Sopra a tutt*i* e' due questi vasi adatterai [uno] canale bene asciutto. E capi abbino ritorti e di sotto presso al fondo dell'uno e dell'altro vaso. Abbi il vaso di piombo la bocca un poco di sopra, cioè ritta acciò che l'acqua con alcuno vaso [in] quello possi infondersi. Faccisi in questo vaso, sotto, un altro canale, breve e piccolo, acciò che inde si vacui il acqua, quando vorrai.

Esempri grazia sia il corno nota "a", *il* vaso simile al bicchiere, nota "b", e la nota del vas*o* [di] piombo sia "g". Canale di quello, piccolo, nota "d". El canne grande commune, nota "e", e la bocca piccola, che ène sopra il vaso di piombo, nota "z".

E'quali tutti, come aviamo detto, adatti, [sia] isturato il canne piccolo di sotto ed empi il vaso di piombo d'acqua per la bocca di sopr*a*. Tura poscia quello, per modo che non esca niente. Infonde ancora il acqua in quello vaso che ène simile allo bicchiere, e secondo la quantità del corno, che in *nel v*aso possa stare e impire. Dapoi apri il canale di *so*tto. Ed escirà [l']acqua. E quando sarà evacuata tutta quella che ène nel vaso, attrarràgli il aria *dal* corno. E quando escirà, tirerà a sè il acqua che *ène* nello vaso.

E quello che avviene è simile *a que*llo di questo canale. Al viso tuo sarà manifesto, *se tu* ci metterai la considerazione e la fantasia, che sicondo la quantità dell'aria che escirà del corno, *si* de[ve] levare la parte dell'acqua, penta coll'arie, potente essa elevare.

---

[343] Here as elsewhere the translation is based on a text identical or similar to **P;** however, distinctive characteristics of the drawing of this chapter, in **P,** are replaced by a tradition more similar to tradition L-I.

[344] **E, P:** subtus vel iuxta. See notes 224, 416.

Elevata, poi discenderà, *per*chè per natura si va. Discende siccome già dimostriamo che l'acqua molle all'arie tenacemente *(fo. 3r)* continua. E secondo la quantità dell'arie che escirà a essa, la parte dell'acqua succederà, e tene[n]te il suo luogo, s'empirà. La quale escente, il arie entrerà, acciò che riempia il suo luogo.

E questa ène *la* figura.

[Chapter 5]

(Fo. 3v) E poi che noi aviamo chiar[i]to questo, è da ricordare uno altro ingegno, simile a quello di sopra, nel quale s'insegnava, che l'acqua il suo moto sempre ène all'in giù, siccome il moto degli altri corpi gravi, siccome noi diciemmo, sicondo il modo naturale. Anco alcuna volta si fa saglire, per impeto veloce, della ragione e però [che] il acqua [sta] in luogo piano e non appiaggia, [sic]chè stanno immobili: Le quali quando alcuno le vorrà co[n] moto natur*ale* elevare a luogo alto,[345] è bisogno d'avere ingegno competenti. La quale cosa a molti ène ignota, che no*n* sanno cavare la detta acqua da detti luoghi, sino colle secchie o con istrumenti che si tiranno d'animali, o veramente qualche rivo più basso. Ma noi sottilemente insegnaremo come egli si possi fare, bene che alcuni si pensano il saperlo e vantanlo, e non sanno niente. Torniamo a proposito.

[Chapter 6]

[a]

Per canale ritorto, che d'alcuni circio si chiama Egittiaco, [l']acqua in nel stagno stante elevanno a luogo alto. La quale poi tanto [d]iscende di sotto, quanto sale. *Non* senza questo ingegno si porrà questa acqua elevare. *Ch*e con questo canale si levi, non è si non per questo ch*e* io dirò.

Quando porremo il capo di questo canale ritorto in vaso d'acqua pieno ed alcuna parte d'a*cqua* surgendo con bocca attrarremo, si surgerà quel*la per* *t*rarre il alito, come aviamo detto. E poi, quando *co*mincierà ascendere, sarà la sua salita *c*ontinua, per insino che 'l vaso sarà votio. E da poi *per* quello canale il acqua s'accostarà, per la quale *(fo. 4r)* cosa [l'acqua] mai si spartirà, se l'aria non ci si trammette.

Chè se nel canale ci s'intromette la detta aria longo tempo, [ri]muoverà la sua tenacità e l'acqua dividerà segr*e*gata, e quella che rimarrà, starà ferma.

[b]

E questo ène l'esempro. Sia uno vaso longo, asciutto, che posto nell'acqua si colligie in essa [l']acqua, insino ch'egli sia pieno. Si r[o]vis[c]i cautamente, sotto l'acqua rivoltato. Dapoi pianamente e a poco a poco se levi, per insino

[345] **E, P:** "quos cum quis cum motu *innaturali* elevaret:" see note 233.

che qu*a*si tutto il vaso si cavi fuore, nientedimeno la sua bocca instante sotto l'acqua. Questo fatto, sarà questo vaso pieno, bene che rivol*tato*. E che questo sie è chiaro. Se fusse di vetro, si v*e*drebbe chiaramente essere pieno. E non è alcuno ingegno che dello detto vaso ne possi cavare l'acqua, che se nello detto vaso sarà uno buco nel fund*o*. Tanto che l'aria ci possi entrare, il acqua ritornerà in giù, e ritornerà al luogo, là dove era.

È *ma*nifesto adunque, che l'acqua si continua in l'arie, con esso penta, e però sempre sono vicini e l'uno segu*e* l'altro.

[Chapter 7]

(Fo. 4v) La natura del fuoco è misto ancora coll'aria e però il arie con esso s'attrae, come apparà manifestamente di sotto. Fammo adunque una palla di piombo, vuotia dentro e capace e mediocre in grandezza, non molto sottile perchè non si spezzi presto, nè sia porosa, e bene asciutta acciò che si facci meglio quello che noi vogliamo. Da poi sia perforata nella sommità, e mettervi uno canale ritorto, che discenda quasi per insino al fondo. Pone ancora el capo del canale medesimo in uno altro vaso presso, ch'è pieno d'acqua, presso al fond*o* del vaso, come nel primo dissi che tu lo mettessi presso al fondo, acciò che meglio far ci corga. Sia la palla "a" e il canale "b", el vaso pieno "g".

Dico ad*un*que che se porrai la detta palla opposita al sole, quando sarà riscaldata di fuore, n'escirà parte d'aria inclusa nel canale. E questo sarà mani*festo* al vedere nostro, perchè cadrà dal canale in nell'acqua e mescolaràssi con essa e farà assai lanpolle, l'una dopo l'altra. Ma si tu porrai la detta palla all'ombra, a dove i razi [raggi] del sole non sian, discenderà l'acqua per [il] canale, per insino che discenderà nella palla.

Da poi se tu la riporrai al sole, ritornerà il acqua nell'altro vasi, e così per contrario. Quante volte lo farai, così sempre tr[o]vera[i]. E se tu riscald*e*rai detta palla al fuoco, quello medesimo troverai. E se ancora tu la porrai in acqua calda, s'ella no[n] diventa fredda,[346] n[e] uscirà.

[Chapter 8]

(Fo. 5r) Perchè voglio dichiarare che non si dà luogo vacuo.[347] E se *fu*sse vuotia dall'aria, immediate vi succede qualche corpo che si mista con la detta

[346] **P** also writes: "si *non* refrigidata fuerit." Compare note 249 about tradition L-I.

[347] Compare note 250 about the different expressions, proposed by the Latin texts. At this point, relating to basic Aristotelian doctrine, the scribes—including the Italian interpreter—are more ready than usual to "correct" the substance of the statement attributed to Philo, and to make it more Aristotelian, less Democritian.

aria, perchè l'uno pegnie l'altro. E questo affermano e'professori delle iscienze. E per questo non si dà luogo vacuo, che da aria o d'acqua o d'altro corpo non sie pieno.

Del quale questo ène l'esempro. Infondiamo acqua in vaso che sia "a", nel quale in mezzo sia uno appoggio, che sia "b", a modo di uno candeliere, sopra a l'acqua ritto. Ed in esso poniamo una candela accesa, che sia "c", sopra alla quale si converte uno vaso, che sia "d", sicchè la sua bocca sia presso all'acqua. E sia il luogo della candela nel suo mezzo. Questo fatto, uno poco di poi vedrai il acqua corre nel vaso di sotto. Vedrai saglire suso al vaso di sopra.

La quale cosa non avvi*ene* si non per quello che noi ricordamo: perchè l'aria in quello vaso contento periscie per l'accensione del fuoco e non può durare col fuoco. Da poi che ène perita quell'aria per lo moto del fuoco, avviene che se leverà il acqua, sicondo la quantità di quello che ène perito d*e*ll'arie. E questo accidente è simile a quello che avviene al canale predetto, ma in questo vaso della candela sopposto, il arie si consuma ed invecchia. Per lo fumo si consuma. E però e's'eleva, il acqua succendente e su*b*intrante empie il luogo dell'aria, che vacuo diven*i*va. E questo ène la figura.

[Chapter 9]

(Fo. 5v) È manifesto per le cose dette che, posta una parte del canale in vaso pieno d'acqua, chi suchiarà dall'altra, cavarà in ogni modo l'acqua, per insino che si vuoti con quello canale. Unde poniamo questo esempro.

Bucaràssi il vaso "a" ne*i* lati presso alla sommità "b" e ponghici il canale "c" per insino appresso al fondo del vaso. Fermamente nel forame si leghi attato col vaso. E sia la parte del canale di fuore alquanto longa. Discenda all'in giù, più che quella di dentro.

Da poi empisi il vaso d'acqua, per insino che s'elevi alla sommità del canale. E cominciarà anda*re*, perchè l'acqua ène più alta e quando s'empie la parte interiore "c" e l'addizione dell'acqua pegnie i*l a*ria, che ène in "c". E in esso pegniare non ce*ssa* d'andare per insino che sia votio, benchè l'acqua sia nel vaso. E questo ène la figura.

[Chapter 10][348]

(Fo. 6r) Alcuni maestri di questi artifizi sforzati si sono per più fa*cil*mente fare una generazione di vasi, dentro nasconden*ti*, ponenti el canale in questo modo. Sia uno vaso "abc", forato di sotto nel mezzo. A dove ène il "c", intromettesi uno canale per lo buco e ine istìa fermamente addattato, che sia "cd", drittamente dritto quasi alla sommità di "ab". A qu*ello* s'adatti uno altro

[348] **P** has a very confused drawing here. **I** clarifies it to some extent.

canale, etto a vaso fermato, che sia "ctk". Sia nientedimeno, che las*ciansi* al "tk" forati, acciò che l'acqua possi d'inde intrare, s*e i* vasi saranno messi infusi.[349] A "k" *quando ascende*, inverso perverrà per "dg" descendente e guazzante. E questo ène il vaso, in giù stante fatto.[350]

[Chapter 11]

Dopo questo è da dire questo. Fa uno vaso, che sia "ab", nel quale fondo sieno buchi minuti, dove sieno "ch". E quando che tu empirai il detto vaso per impito nell'acqua, si sturarai la bocca del vaso col dito, niente uscirà. Ma quando tu levarai il dito della bocca, cominciarà il acqua a uscire.

Ma guarda che'l vas*o* non sia di troppa capacità per la ponderosità dell'acqua, perchè s'ella fusse troppa, discenderebbe per forza.

[Chapter 12]

Chè se a questo vaso pieno una foglia o uno foglio o simile a quello alla bocca apporrai e colla mano alla bocca que*lla* accostarai, accostandolo [dri]tto al vaso, resterà ine la foglia o'l foglio, pendente in quello medesimo modo, sicc*o*me egli ci fusse confitto.

O ancora se ci accostarai uno crivello alla bocca, niente d'acqua n'escirà.

La ragione è questa che di qua e di là il acqua correndo per gli buchi minuti, gli impirà, nè più discenderà, perchè non può il aria entrare, cioè ponendo il detto crivello sotto sopra, perchè non si dà vacua, non avendo il aria dove entrare. Il acqua rimane e sta lì e non si parte "b".

[Chapter 17]

(Fo. 6v) Faccisi ancora uno vaso colla bocca stretta, e sia "ab" e sia perforato in fondo, dove sia "d". Per quale entri per lo dritto uno canale, in sommo ritorto [e] gionghi presso al collo del vaso, che sia "cde". E sia la parte, che ène nel vaso, ritta e l'altra parte, la quale ène "cd", sia di fuore stesa.[351] Della quale l'estremità, la quale sia "[d]e", entri in uno altro vaso, che sia "chz", e sia sott*o*forato. Ed adattansi questi due vasi, che sieno sopra un certo concavo che sia "mn", sicchè el acqua, che esce del fondo del vaso "chz"

---

[349] Latin: "aqua possit ibidem introire, si vasi fuerit infusa." The Anonymus misinterpretes "vasi" as a nominative plural.

[350] See note 273. The text of the Anonymus is rather original.

[351] Here, as in Chapters 8–11 and 18–21, the "g" of the Latin texts (including **P**) is rendered as "c". In Chapters 4 and 7, "g" is retained. Perhaps the Anonymus tried to "Latinize the Latin" while translating it into Italian.

nel luogo "c", possi entrare, e non altrui. Adattisi ancora uno altro canale nel vaso "[a]b", escente nel luogo "t", disteso sì che il suo capo, che sia "p", se levi sopra il vaso "chz". El quale canale sia "pqr".

Quando tu empirai il vaso d'acqua, "ab", meno d'altitu*di*ne "de", e sturerai la bocca sua [con] el cuperchio, che sia "xy", non cesserà il acqua d'uscire nel vaso "chz", per insino che sarà pieno, sopra a quello che si ferma al concavo.[352] Discenderà il acqua per insino al "c". El quale, d'acqua ripieno, non a[v]rà aria che entri nel vaso "ab". E però cesserà il acqua di correre per "p" per insino che si vuoti in parte "c". E allora escirà, come prima. E così istarà sempre nella medesima altitudine. E questa ène la figura.

[Chapter 21][353]

(Fo. 7r) Ad idem

Fa*remo* uno vaso forato [di] sotto, avente il canale di sotto, che sia "c", da qualunque parte [in] nove o più parti distinto. El quale sia "abbccdd". E de'quali l'uno e l'altro, e sopra e sotto, sia forato. [D]e'quali, tutti d'ogni licore ripieni, se [gli] forami di sopra sturerai, niente uscirà per di sotto. E [di] qualunque di sopra aprirai, discenderà il licore per di sotto, escendo per lo canale del vaso, che ène "c", come manifesta la presente figura.

[Chapter 18]

Ad idem

Si fa un altro vaso della medesima generazione, ma più leggero, e sia "a[b]" e sia sopra due colonne "d" ed "e". Del quale la bocca sia "c", el canale che esce "rp". Sia una delle colonne concava. Per la quale subintri nel vaso "ab" el canne estenso al vaso "ghz" e subintrante nel luogo "z", che sia il canale "qfz". Adatti*si* uno vaso "ghz" supposto alla bocca del canne [ed] entr*i* solo per "p", sicchè il acqua, che ène escente a "p", vada in esso. Impito adunque il vaso "ab" sicond*o* l'altitudine "c" e sturata la bocca sua, che de' . . ., escirà l'acqua per "p", ancora nel vaso "ghz" . . ., sino che s'empie . . . E sarà in ogni modo questo vaso come nel predetto. E questa è la figura.

---

352 In this sentence and the next, the Anonymus follows the Latin text, although it fails to make any technical sense in view of the drawing. His attempt to critically renew the tradition meets aith difficulties.

353 Inserted before Chapter 18, in substantial accordance with tradition L-II. The final clause of this chapter, "come manifesta la presente figura," became ubiquitous in the Trattati of Martini.

[Chapter 19]

(Fo. 7v) Ad idem

*Fa*ncisi altri vasi "ab", "ghz", co'suoi canali "pqr", "cde", in nel medesimo modo come prima, sino che s'interponga panni,[354] che sian "st". Impito d'acqua il vaso ["ab"], quasi per insino al "p", e la sua bocca isturatta, che sia "o", descenderà l'acqua per "cd", escente da "c" e d'inde subintrante nel vaso "ghz", per insino che pervenga [da "r"]. E così sarà come prima.

[Chapter 20]

Pel medesimo faccisi uno vaso, d'una superficie, che sia "abc". Abbi da'lati due canali escenti de esso, che sieno "cd", "be". E abbi ancora uno canale dentro, ritto, discendente nel vaso "ghz". E fermamente ciascuno de' vasi istìa ne'loro luoghi "nm", che sia il canale "klmn". Siano certe particule del vaso "ghz", a modo di crogioli, da'lati escenti e scese da'canali "cd", "be", e le dette particule sieno "gt", "zs".

(Fo. 8r) E quando adunque empirai il vaso "abc", meno dell'altitudine "n", escirà il licore che ci metti, per "cd" in "sz" e per "be" in "tg", qua e là andante nel vaso, [per] "k", el quale empi*e*, *e* cesserà il venire [a] "d" e "b".

Sia il licore nel vaso "abc" cioè olio ed adattesi uno taccolo nel vaso "ghz", dove egli ène *l'olio*. E sicondo che ardendo si consumerà dell'*olio di* "ghz", a poco a poco discenderà da "abc" per "d" ed ["e"].

Questi nove[355] sono *del*la medesima generazione. E essa è la finis.

---

[354] The drawing shows a column with a ribbon wound around it.

[355] The Anonymus mistakes a Latin "IV" (see note 301) for IX.

The Arabic *Pneumatica* and its Latin Supplements

The annotated, English text of Philo's *Pneumatica*, which follows here, renders the contents of the Arabic manuscript, A.S. 3713 (**A**), supplemented by the Latin Philo (**C** to **Z**). It translates everything as literally as idiomatic English allows. The English text generally follows the French one of Carra de Vaux, which has been found most accurate.[401]

A few errors of the Carra de Vaux translation have been found in the intervening time. They are corrected here, as mentioned in footnotes. As shown by this annotation, some corrections are based on the authentic drawings; some are developed from a technical review; and some simply reflect the re-reading of the text by Mrs. Showker and by Mr. and Mrs. Keller. This latter re-reading confirmed the opinion of Drachmann—a most qualified judge—, who had found the French translation of Carra de Vaux a most accurate restatement of the Arabic text, expressive even of small nuances of the wording.[401] The very few exceptions from this rule, found while preparing this edition, are identified in the notes.

The format of the authentic pages in manuscript **A** is 13 by 17 cm. The Arabic text and the Arabic reference letters and legends on the drawings are written by a single, generally very clear hand. A chapter begins with an introductory statement, written in heavy, colored ink. A chapter ends with a drawing. This is followed—often on the same page—by the next chapter-introduction.

In the present translation, as in the French one of Carra de Vaux, the reference letters are re-translated into their Greek form, for the convenience of the western reader. The specific correlation of the Greek and Arabic letters (and sometimes numerals) is shown in explanations of the drawings.

These drawings are photographed here from manuscript **A** and offset-printed from the photographic film. The drawings of the Latin supplements and variants will be found in the preceding facsimile of manuscript **N**, under the corresponding chapter numbers.

The reference letters and legends on the drawings of **A** are explained beneath the reproductions of the first two scores of illustrations. From then on, speci-

[401] Drachmann, p. 42. Somewhat more guarded: Sarton, II, p. 347; Wiedemann-Hauser on AlJazari, Water Lifters, p. 122n.

fically from Chapter 44 on, the explanation is omitted, in general because we found the legends undecipherable, as in Chapter 58 (Figure 58), or because we found them difficult and the reference letters obvious. In translating the legends we used the following system:

If the legend appears on the drawing itself, with or without reference letter, we show its translation between quotation marks; so for example for Chapters 2, 6, 7, 8, 13 (Figures 11–14, 17).

In most cases, where the explanation of the reference letters is given only in the Arabic text, not on its drawing, we show this explanation between brackets and quotation marks; so for example in the first three references of the drawing of Chapter 11 (figure 16).

Finally, in a few cases, where we add an editorial explanation, we show it between brackets, without quotation marks, as in some of the explanations for Chapters 10, 15 to 17 . . . (Figures 15, 19 to 21 . . .).

The contents of each chapter are annotated here in an attempt to clarify their scientific, technical and historic problems. This annotation makes no attempt to compare A. S. 3713 with the later, Arabic copy book and extract book, A. S. 2755 and Eastern 954. The annotation, however, traces the designs back to the earlier versions, presently known: mainly to Ahmed Ibn Musa and to Hero, and in some respects to archaeological findings.

This annotation also suggests a few conclusions about the original form of Philo's own *Pneumatica*. These latter notes, of course, must be understood as being hypothetical. About four centuries elapsed between the times of Philo and Hero; about seven from then to Ahmed; and another five from Ahmed to A. S. 3713. Too little is known, today, about those centuries to allow anything but a hypothesis about their history of technical science.

Briefly, the present hypothesis is as follows. A. S. 3713 is a copy of a rather original version of Philo's *Pneumatica*. The initial score of chapters, stating and demonstrating Philo's theory, is shown more literally in the Latin version. Three of the other 42 chapters (55 to 57) are also known in a Latin-Gothic set of fragments. The entire text was translated into Persian. Then an Arabic prototype originated, using the Arabic variant of AlJazari (Chapter 30), western sources (Chapters 18 and 33–36), and other traditions (mainly in Chapters 51 to 65). In the next century, A. S. 3713 was copied from the prototype.

English Version, based on A. S 3713, with historical notes

In the Name of God, the Merciful and Forgiving:
The Book of Philo on Instruments of
Pneumatics and Machines for Water.

[Chapter 1. Introduction. Air is a Body, not a void][402]

Dear Muristom: Your interest in ingenious devices has been known to me. You say and urge that you want a book about them. I wrote it and send it glady. May it be an aid to your studies of all these devices. Indeed such matters are worth the attention of learned men.

Natural philosophers, denying a common opinion, have said that a vessel never is empty, though it may so appear, but is full of some things;[403] men had not recognized that air is one of these things. I will discuss this briefly, not citing opinions or counter-opinions, to avoid prolixity. It has been said clearly, and can be shown to the senses, that air is an element, a thing, not only a name. Briefly, I will show that it is a tangible thing.

[Chapter 2. Demonstration to Show that Air is a Body: Air-Filled Carafe under Water][404]

Let us take an empty vessel, similar to the carafes made in Egypt, and plunge it in deep water, with its mouth inverted, while pushing it down. Not the least

---

402 NOTE: Here and generally throughout the first score of chapters, the Latin text coincides in substance with the Arabic one, and the present English translation renders the Arabic-Latin *consensus*. Exceptional Latin variants will be translated in footnotes (beginning with the next note), or in text inserts, when relatively large (beginning with Chapter 11).

403 Latin text: "... *a vessel never is empty, though it may so appear* ... *but is full of air or of other things.*" Similarly Hero, Introduction. pp. 4, 5.

404 The Latin and Arabic manuscripts have legends, not reference letters, in their drawings to Chapters 2 and 6. Similarly Hero, Introduction, pp. 1–29. These facts give added support to the conclusion (of Diels, Straton, *passim;* also Gottschalk[24], pp. 127, 141), that the description of the basic experiments—Chapters 2 and 6—is or contains a fragment of Strato's book *On the Void.*

bit of water enters, until some of the air escapes. Upon the escape of air there follows the entry of water. Which I can demonstrate by this experiment.

Let us take an Egyptian carafe with a narrow mouth. Let its bottom have a minute hole, and let it be closed with wax. Then let it be turned over, and let it be put in deep water, taking care to hold it straight and not inclined to either

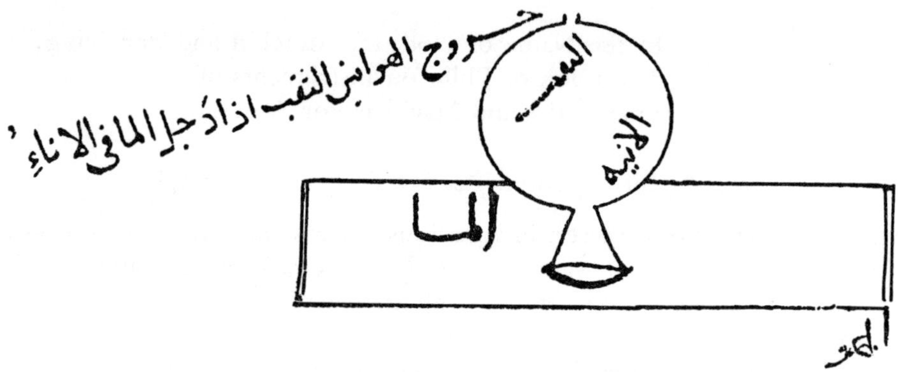

Figure 11. A. S. 3713, fo. 3a (bottom). *Drawing of Chapter 2.* Legends:
at lower right: "The Air"
in the rectangle: "The Water"
in the circle, lower right: "The Container"
upper left: "The Hole"
above the circle: "Exit of Air from the Hole, as the water enters below"

side. Let it be pushed down in the water and then taken out slowly and gently. The inside will be found dry. It is wetted only on the mouthpiece.[405] Thus it is clear that air is a body. If it were not a body, and the space inside were void, water would have run into it, as it would have found no resistance.[406]

[405] Latin text: "*Let us take a vessel as aforesaid, with a narrow mouth, . . . reverse it, mouth down, and . . . push it down with our hands to submerge it entirely in water. . . . When it is lifted, it will be found . . . wetted only at the outer part of the mouthpiece.*" Hero's account, Introduction, end of page 5, is similar to the Latin version. Also see Hammer-Jensen, pp. 414f.; Drachmann, p. 47. Archaic devices, similar to that of Chapter 2 but more complex, may be found in Figure 2a and f. — "Egyptian" vessels and other "Egyptian" devices (see Index) are mentioned in the Latin and Arabic traditions. Evidently they were mentioned in the archetype, and probably in Philo's original.

[406] Aristotle had mentioned that someone had, qualitatively, observed that air has weight: *De coelo*, IV, 4, p. 311b 10; see Gerland-Traumüller[79], pp. 268f., also Galileo, *Discorso*[79], p. 122. Hero as well as Philo is silent about this fact, so far as his *Pneumatica* is preserved. The problem is briefly discussed by Cohen-Drabkin[89], p. 247f.

Again, let the aforesaid vessel be inverted, and manually submerged in water for some time. Then let the wax, placed on the hole, be removed. There then ensues an escape of air from the hole, which is manifest to the senses. One can see it in form of bubbles above the hole. The vessel then fills with water, as air leaves through the hole. What forces the air to escape is the motion of the water, which presses the air while we push the vessel down. The demonstration shows that air is a body. Here is the drawing.

[Chapter 3. Atoms, Voids, and Cohesion][407]

We shall now speak of other elements, which are important for instruction in this science.

Philosophers have said that air consists of very minute particles, which due to their smallness cannot be comprehended by the sense of sight or any other sense, when they are separated, and that the air can then be noted only by its force.[408] When they are joined, it is not so.

A philosopher also believes that a void is part of nature and that it is mingled with the body of air, and enters among its light particles, as it also does among the particles of liquids and of other things. This we have explained at another point, where we wrote "De Mirabilibus Arbitriis."[409]

---

[407] Many opinions have been stated about the contents of this chapter, the identities of the "philosophers" cited in its second paragraph and of the "one philosopher" cited in the Latin version of the third paragraph, and the nature of Philo's *De mirabilibus arbitriis.* See, for example, Crombie[176], II, pp. 37–40 and Krafft, pp. 368, 372. In my opinion the context shows that paragraph 2 of Chapter 3 refers to Democritus; Latin paragraph 3 refers to Strato *(a reference that was lost in the independent, Arabic translation)*; and the cited book or chapter was part of Philo's *Introduction* or of the original *Pneumatica.* The view expressed in Hero's Introduction, pp. 6 and 7, coincides with the present paragraphs 2, 3.

[408] Latin text: "*they* [the particles of air] *cannot then* [when separated] *be noted, except by error.* "*Evidently the Latin, Philonian tradition about this important problem of pneumatic physics differed from the Arabic, Philonian tradition.*

[409] Latin text: "*One philosopher also says . . .*" Another disagreement between the traditions. *These disagreements are disregarded by the current theory about the derivation of the Latin text,* and by many current hypotheses about the interpretation of either text. The word *arbitria* and its Arabic equivalent is interpreted by Schmidt, p. 465, and by Carra de Vaux, p. 37, as a corrupted form of "Automata" or of musical "Organs". However, the text clearly shows that it was an expression for the doctrine of atoms and vacua in the elements.

These men believe that the substances of the liquid element and of air are combined by nature, and the two leave no distance between them.[410]

Therefore it happens often that water travels upward. However, its weight, which naturally prevails, then forces it down, for such is the property of all heavy bodies that by their nature they tend downwards.[411]

[Chapter 4. Water can be drawn up by Rising Air. The Wine Taster and the Interrupted Siphon]

[a]

So it is manifest that water sometimes rises because it is drawn up with rising air, which is connected with it. This is clear from the operation which occurs with the tube used for wine tasting.[412] When someone holds its tip in his mouth and gently sucks out the air that is in it, he also attracts with this air the liquid body which is down below, since it adheres to the air as if it were held to it with glue or with some other bond.

[b][413]

This can also be shown as follows. Let there be a bull's horn, well hollowed out inside to make it wide, thin and polished. Let it be of medium height. Let its form be round, with its "lower" end [frusto-]conically shaped.[414]

---

410 Latin text: "*Let us now turn to the experiment. It may be said that the liquid and airy elements are combined by a bond that is more tenaceous than others, and there is no distance between them.*"

411 Hero's Introduction (pp. 8, 9), based on Strato and Philo, adds here: "When any force is applied to it, the air is compressed and, contrary to its nature, moves into the vacant spaces, from the pressure applied to its particles; when the force is withdrawn, the air returns to its former position, from the elasticity *(eutonia)* of its particles . . . Similarly if . . . the particles be divided and a vacuum produced larger than is natural . . ."

412 One manuscript, **F**, has an illustration of this device: a short, straight tube, obliquely inserted in a full cup. Hero describes the process without illustration: Introduction, pp. 10f.

413 The "interrupted siphon." A variant is shown by Hero, II, 14. Both men are silent about the fact that the "small pipe," used as outlet of the second vessel or "can", must go down substantially below the water level: a fact explicitly stated only about 1600 by Della Porta, pp. 35f. Also see Schmidt, XVII. Jahrhundert, pp. 207f.; Drachmann, pp. 47f., 50.

414 Meaning: let its wide end be ground to tapering form to fit a wooden lid and adaptor, for connection to the upper, connecting pipe. The wide end of the horn (originally attached to the ox-skull) is still called its "lower" end: an expression

Around the cone-shaped part let a wooden lid be tightly fitted, to achieve the construction that we need, [that is,] let them be so conformed that air cannot escape anywhere and the entire horn unit resembles a medicine can.

[Latin Supplement:

Then invert this horn and let it stand in some wide-mouthed vessel similar to a beaker.[415]

Place another can, made of lead, at a level below and also alongside this one.[416] *Let it be well sealed so that, when necessary, air may not pass it.*

Between the tops of these two cans adapt a pipe, very tight, with inverted end pieces going down to the bottom [to the end?] of either can. Let the lead can have a mouth, a little raised, so that water can be poured in from some vessel. Let this can have a lower pipe, well attached and small, so that water can be evacuated when you wish. Let the horn, for example, be called "a", the beaker "b", the lead can "g", the small pipe thereof "d", large common pipe "e" and the small mouth on top of the lead can "z". Arrange all this as we have said.[417]

Then close the small pipe below; fill the lead can with water through the mouth on top; close that mouth so that nothing will pass; pour water into the beaker-shaped vessel, in such amount that it can fill the horn standing in it. Then open the closed lower pipe, and water will come from it. As it flows freely from the lead can, air is attracted [into the can] from the horn. As the air leaves there, it draws with it water from the beaker [into the horn]. What enters here replaces what leaves through the pipe. You may see it [through the horn] if you give it due consideration. According to the amount of air leaving the horn, a portion of water is raised, being held to the air, which can lift

---

which seems to have misled most of the illustrators, other than the draftsman of manuscript **C.**

415 Cyphus (**D, V**); ciphus (in **C, E, G, H** as well as **N;** also in **P, R, S, W, X, Y,** and **Z**); cifus (**F**); bicchiere (**I**); scyphus in classic Latin; skyphos in Greek. *The Latin expression seems to be derived, with some corruption, directly from the Greek archetype; it has no equivalent in Arabic.*

416 **F:** "subtus at iuxta." **I:** "sotto." Other Latin manuscripts: "subtus vel iuxta." The terms used in **F** and **I** give an imperfect reflection, but a better one than the other term, of the physical construction which is required according to the modern pneumatics, as first described by della Porta.[413] The texts of Philo and Hero make the impression that the device had been demonstrated (perhaps by Ctesibius) but that it was not fully understood.

417 Possibly referring to oral explanations about the critical point, mentioned in notes 413 and 416.

it. After such lifting it descends, for its nature has this effect, as we have demonstrated.[418] *The fluid water follows the air tenaceously. As an amount of air leaves, a portion of water follows and occupies its place, filling it; and when such water leaves, air enters to refill its place.*]

[Chapter 5. Impetus of the Drawing up of Water. Water-Lifting Techniques: Cognoscenti and Ignoramuses][419]

[Latin Text:

As we have clarified this, a device is to be recorded which resembles the preceding one, whereby it will be taught that by nature the motion of water also leads down, as does the motion of the other heavy bodies, which we have mentioned. An artificial motion, or rather a natural-artificial one,[420] *sometimes raises it to some extent. It can be enhanced by an energetic drawing-up effect.*[421] *When water stands quiet and immobile in a plane flat place, and when one wishes artificially to raise it to a high place, one needs a powerful device. Such device is unknown to certain people, who do not know how to take water from such places, except with buckets, as in wells, or with devices moved and drawn by animals, or in some cases by extraction, using streams from a river or well, which descend to lower levels. Hereafter, we will show how it can be done much more elegantly.*[422] *No doubt there are certain people who think and announce they know this device, and who will brag about books they will compose about it,*[423] *when (actually), they have been quite ignorant about it. They deceive themselves and their listeners.*[424] *Enough of this, let us come to our point.*]

---

[418] This is only *mentioned* in the preceding Chapter 3; the process is "demonstrated" only in Chapters 5, 6. Hammer-Jensen, p. 489, plausibly concludes that Chapter 4(b) belongs to a position after Chapters 5. 6.

[419] The *lacuna* of the Arabic version, which began shortly after the beginning of Chapter 4b, continues here.

[420] Interpreting the word "medius" of the Latin text as meaning "between natural and unnatural." The corrections proposed by De Rochas and Schmidt ("modicus" or "meditatus") do not yield any technical sense.

[421] "*Impetus attractionis.*" The term "*impetus*" (Greek: ropē) had been used by Aristotle, for example *De coelo* IV, 1, p. 308a1. As is well known, the concept was changed and developed by Simplicius, Philoponus, Buridanus and others.

[422] In a papyrus ascribed to the III. century, the writer — conceivably either Philo or his rival — appeals to the king about a wonderful water-raising device he has invented, and asks to be attached to a "strategus Ariston." Private information of Thorkild Schiøler, received through Alex Keller.

[423] Whether such books were actually composed is unknown.

[424] Similar debate arose when medieval promotors proposed mountain-spanning siphons, to supply mills or cities with water. The debate is reflected by a remark of

[Chapter 6. Drawing-up Effect (Continued). Siphon. Water-Lifting Beaker]

[a]

Take a bent pipe, which is sometimes called Egyptian Compass.[425] *It raises water, which stands in a pool, to a high place. The water then, however, descends as far as it has risen before it was raised with this device.* Only this way can water be raised [with it].[426] When the water is raised with this device, it is only for the reason which we will show.

When we put the end of this bent pipe in a vessel full of water, and attract some part of this water by sucking it out with our mouth, the water rises, as we have said. When it once begins to ascend, the outflow is continuous until the vessel is empty. The water coheres with its preceding parts in this pipe. It never separates, unless air be introduced.

If the siphon breaks and air enters the pipe, this interrupts the water's continuity. That which is still there then stands still,[427] for the reason that we have indicated.

[b]

The following experiment is a demonstration for it. Let there be a long, tight vessel. Let it be put in water and pressed down until it is full. When it has thus been filled, overturn it quickly under the water, while keeping it full. Then lift it from the water until only its lip stays under water. When

---

Brunelleschi, see note 332. Also see Taccola, *De ingeneis*[77], fo. 73v; same, *De machinis*[78], fo. 50v, 51r; Kyeser[3], fo. 61v; Martini[3], pl. 48, 74f., 78, 81; Leonardo, *Les manuscrits*, ed. C. Ravaisson-Mollien, Paris, 1881–91, vol. B, fo. 26r; Della Porta[78], pp. 33, 63. The unworkable nature of this proposal was noted by Galileo, and explained by Torricelli; thereafter Pascal—following a trend begun by Della Porta and others—, blamed Hero with the unworkable proposal: *Traitez de l'équilibre*, Paris, 1663 (P. 460 in ed. J. Chevalier, Paris, 1954). Pascal's view was echoed by Diderot[83], XV, p. 765. The fact is that the Latin Philo has more to do with this proposal than Hero.

425 Greek: *diabetes*. Compare Hero, I, 3.

426 The Latin text (which had to be translated up to this point), as well as the Arabic one (which resumes here), implies that "without *this [device]*, the water cannot be raised." This evidently contradicts a number of other and more plausible statements. Therefore we may assume that according to Philo's original statement at the present point, the water cannot be raised by the siphon "without *this descending*." The archetype seems to have corrupted this plausible and consistent statement.

427 Latin: *quieta manente*. Evidently, another error of the archetype was accepted by eastern and western translators. As shown by the "following experiment," the water contained in the siphon does *not* "stand still" any more, after the venting of the siphon.

this has been done, the vessel still is full. That such is the case is readily seen if it is of glass or horn or other transparent matter. There is no means [for this vessel to "exhale" the water], unless the vessel has a small hole so that air can enter it; the water then returns down to the place where it had been.

Thus it is manifest that the water follows the air and is held with it. They mutually follow one another.

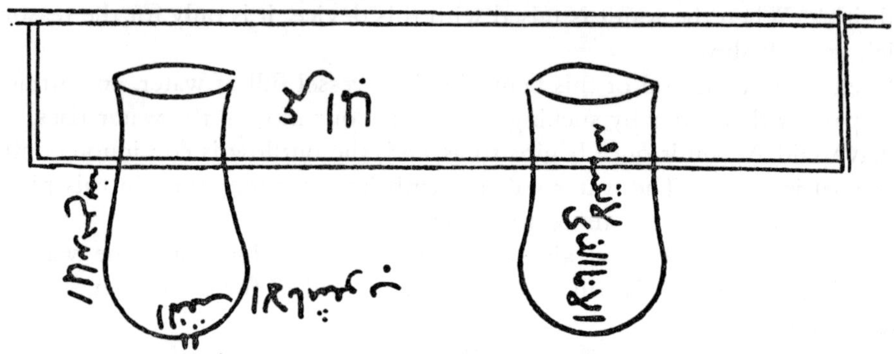

Figure 12. A. S. 3713, fo. 5a (bottom). *Drawing of Chapter 6.* Legends (poorly legible):
at right: "The Water . . . [can enter this vessel?]"
middle, in rectangle: "[? Depth of] . . . the Water"
at left: "Container [? without a Hole]"

[Chapter 7. Water-Lifting Effects of "Fire" and Air. Thermoscopel].[428]

Fire also by nature mixes with air; therefore air is drawn along with it. This will appear clearly from the following.

Let us take an egg of lead, medium sized, hollow and not too thin, so that it may not break too easily. It must be tight, to let us achieve what we want. Let it be drilled, and let a bent pipe be introduced in the hole, which pipe goes down almost to the bottom, so that water may be removed; the pipe must be very tight. The egg is placed in the sunshine. Put a cup below the other end of the pipe. Let the egg be α, the pipe β, and the cup γ.

[428] Hero, II, 8, like the Arabic draftsman and the Latin Philo, uses a "globe" or "ball" (of unspecified material) instead of the "egg" (of lead) used according to the Arabic text. — In later times, Galileo and others claimed invention of the thermometer. R. Fludd, *Philosophia Moysaica*, Gouda, 1638, fols. 1, 2, correctly opposed the claims, having found the construction in an old manuscript (probably **D**). See Drachmann, pp. 48f.; Crombie[176], II, p. 260; W. E. K. Middleton, *A History of the Thermometer*, Baltimore, 1966, pp. 5–23, and mainly Schmidt, Thermoskop, *passim*.

I say: when the egg is heated from the outside, part of the included air escapes through the pipe. This is manifest to sight because air comes from the pipe into the water [in the cup], agitates it, and makes many bubbles, one after the other. When you then place the egg in the shade for some time, water ascends from the cup and enters the egg. When you remove the shade and

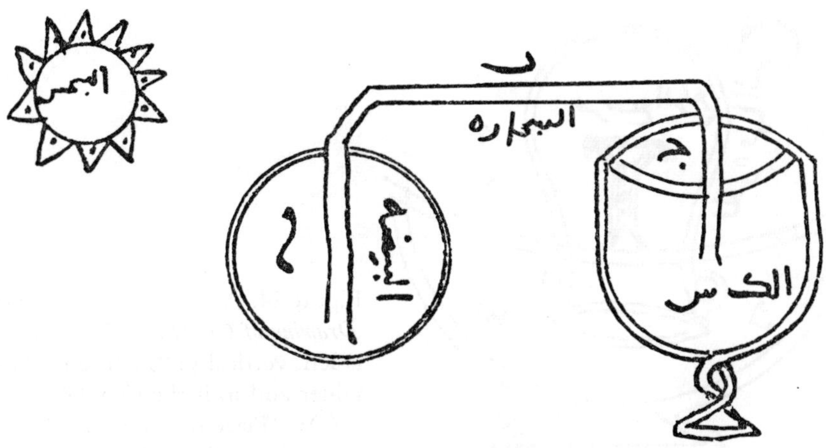

Figure 13. A. S. 3713, fo. 6a (bottom). *Drawing of Chapter 7.* Legends: at left: "The Sun". With Reference Letters:
ا (α): "The Egg"
ب (β): "The Pipe"
ج (γ): "The Cup"

place the egg in the sun again, the water returns into the cup. With the egg in the shade again, the water re-enters it. This goes on, back and forth, as often as you wish.

If you light a fire and bring it to the egg to heat it, the same thing happens, and if the egg then cools, the water returns as before. So also if you put hot water on the egg. Here is the figure.

What we explain here is one of the foundations of the science called pneumatics, which is based on such devices. It occurs only because there can be no void. When air is removed, one of the other bodies, coherent with it, takes its place. They are simply pushed naturally. Such is the opinion of several teachers of natural science, and we agree.[429]

---

[429] In the Arabic text, this paragraph can be read as part of Chapter 8 (where Schmidt plausibly inserts the Latin version, beginning: "What I *will* explain *now*...")

[Chapter 8. Water-Lifting Effect of Fire, Consuming Air: The Water-Lifting Candle].[430]

Pour water in a vessel α, which in its middle has a stand β in form of a candle-holder. On its top we put a candle γ. Above it we invert a flask δ so that its mouth is close to [and below the level of] the water. We let the place

Figure 14. A. S. 3713, fo. 7a (middle). *Drawing of Chapter 8.* Legends: at left, vertically: "Picture of the Container and in it the Candle"
ا (α): "Place of the Candle"
د (δ): "The Water"

of the candle be in the middle of the flask. When this has been done, after a short while you see water of the lower vessel rise up into the flask.[431]

This happens only for the reason that we have noted, namely, that air contained in this vessel perishes and disappears by the burning of the flame

---

I left the paragraph in Chapter 7, where Carra de Vaux has it, for convenience in comparing the English and French translations.

[430] Hero's Introduction, pp. 10f., speaks of "cupping glasses." Either such a glass or the thermoscope process was also mentioned by Aristotle, *De coelo*, IV, 5, p. 312a 5–10: "The water is drawn into the vessel that has been fired." Philo seems to be the first who distinguishes the vessel fired outside (Chapter 7) from that fired inside (Chapter 8). As to this latter one, the next attempts seem to be those of Leonardo (*Codice di Lord Leicester*, ed. Calvi, Milan, 1909, fo. 3v; also see G. Goldbeck, "Die Feuermaschine von Leonardo . . .," *Technikgeschichte*, XXXIX, 1972, pp. 1–10) and of Van Helmont: De Rochas, 1881, p. 17; Schmidt on Leonardo, p. 185; also see J. Beaujeu in R. Taton, *History of Science*, New York, 1963, I, pp. 332f.

[431] The actual process is more complex. Generally it requires preheating of the glasses. Otherwise, water is driven downwards rather than upwards. Also needed is a torch or candle which goes out only when most of the atmospheric oxygen has been used up. As to ancient torches and candles, see Daremberg-Saglio *s. v.* candela, fax.

and cannot endure with it. When air has perished, the motion of the fire causes water to be lifted, in such quantity as air has perished. This effect resembles that which occurs in the aforesaid pipe.[432] Air is removed, dissolved. For this reason the water rises and fills the space which has been emptied. Here is the figure.[433]

[Chapter 9. Siphon Process and Impulse for Starting it].[434]

We have shown [in Chapter 6a] that, when one leg of a two-legged pipe is placed in a vessel full of water, and when someone sucks at the other leg and then stops when he feels the water, the water is drawn out through this pipe,

[Latin Supplement:

until it is entirely exhausted. Whereof we may propose this demonstration. Let vessel "a" be apertured in its wall near the top "b" of its side and let pipe "c" be inserted there, which goes to a point close to the bottom of the vessel. Let it be fitted firmly into the aperture and fitted to the vessel. Let the outer part of the pipe be a little longer, to go down beyond the inner one.

Let the vessel then be filled with water, until it reaches above the top of the pipe. It then begins to run as the water rises higher. It fills the inner part of "c", and the addition of water impels the air in "c". Upon this impulse it continues uninterruptedly to run, until such water as was in the vessel has been emptied.]

[Chapter 10. Concentric Siphon].[435]

Some makers of such devices have tried to improve the construction of one type of vessel by placing a concealed pipe inside, as will be described. It uses an outer tube,

---

[432] It is not clear what "pipe" was meant in Philo's original text. The opposite process occurs in the pipe of Chapter 7: when fire is applied, air will expand, not shrink. The "reason that we have noted" is not stated in the existing book, but may have been recorded in the *De mirabilibus arbitriis*.

[433] The Arabic drawing shows the candle holder upside down. Its reference letters disagree with those of the text. *The Latin Philo is free from these confusions.*

[434] Hero shows different siphon starters, using fitted and extendible pipes, *smerismatia*: I, 6; I, 27; II, 25. However, he knows Philo's starter and uses it as an element: I, 14.

[435] Hero shows and describes substantially the same device: I, 3. He states that "the air which enters the vessel takes the place of the water." The Latin Philo is silent about this point. The Latin illustration and the Arabic text and drawing erroneously call for a vessel with a closed mouth. This is one of the points where the Latin text is relatively closer to Hero, and to proper pneumatics than the Arabic

which must be relatively large in order that the aspiration may not be too violent. The top of this tube must be tightly closed.

Let there be a vessel, perforated in the middle of its bottom. Let a pipe be inserted in the aperture and be firmly fitted there, going vertically up, almost to the top. Let the other tube be adapted around it and secured to the vessel. This tube must be fairly large to avoid all too violent aspiration, and must be tightly closed on top. It must have apertures at the bottom, at both sides, so that the water can enter there, while the tube can stand on the bottom of the vessel and thus the water can flow out easily. The vessel is α; the upright pipe in it β; the outer tube γ; the place which the water reaches δ; and the outlet ε.

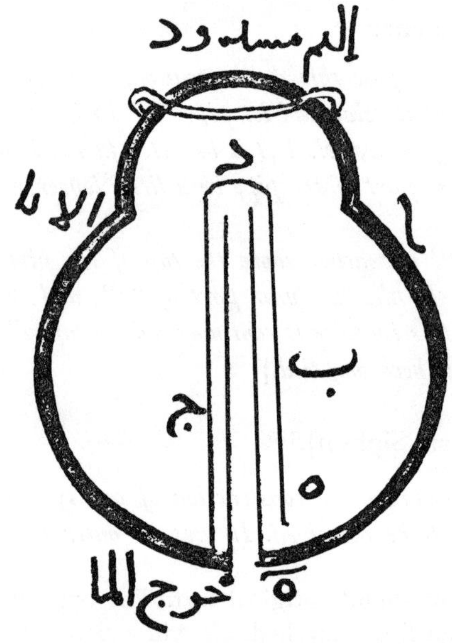

Figure 15. A.S. 3712, fo. 8a (middle). *Drawing of Chapter 10.* Legends and Reference Letters:
top: "The Closed Mouth"
ا (α): "Vessel"
ب (β): [Inner Pipe: "upright pipe"]
ج (γ): [Outer Pipe: "outer tube"]
د (δ): [Siphon Level: "the place which the water reaches"]
ه, ة (ε): "Outlet of the Water"

When this has been prepared, water is poured into the vessel. While this is done, water does not immediately flow out, as the inner pipe rises to a point above the water level. The unit is called a siphon. When the water gradually rises to its top, it [suddenly] descends and emerges. At the beginning of the outflow, the air in the siphon is pushed back, whereon the water fills the

variant. The corresponding part of Ibn Musa's variant—its Chapter 1—also avoids the erroneous call for a closed vessel. NOTE: the *lacuna* of the Arabic text, filled here by the Latin text, ends near the middle of the second paragraph. From there on, the translation again renders the consensus of both texts.

siphon. The water then continues to flow, evidently as in the preceding siphon.[436]

The present siphon is called intermittent.[437] It serves for many experiments, as noted before. It is another foundation of this art. Here is the figure:

[Chapter 11. Water-Lifting Sieve (Clepsydra).][438]

Let us construct another pneumatic vessel and device, fundamental to this science.

Let us take an egg of copper or silver or other material, as desired by the maker. It should be hollow, should hold ½ *qist*, and be tight all around.[439] Drill holes in its surface and insert through them a small pipe, with a clear diameter equal to ½ finger's width, and 1 cubit length. Let it be perfectly attached to the egg with tin solder, to stay firmly in the holes and pass no air. Opposite this pipe let there be small holes in the egg, close to one another like the holes of a sieve. Let the device be elegant, like those for wine.

To use it, take a cup in your left hand and fill it with pure wine. Then take the egg's pipe and fill the egg by submerging it. It fills with the liquid, which enters through the small holes; the air escapes through the pipe at the other side. The servant then firmly closes the mouth of the egg's pipe with his finger, lifts the egg from the liquid, and elevates the egg in the air above the cup. Nothing flows from it. Only when the servant lifts his finger from the pipe the liquid promptly begins to flow. If he then again closes the opening of the pipe with his finger, the effect is as before.

This is for the reasons stated above. While the egg is in the liquid, liquid enters through the holes, as noted, since air escapes through the pipe; otherwise the egg would not be filled. Once the egg is filled and the pipe is closed with a finger, the liquid stays in the egg, because there can be no space void of air.[440]

---

436 The expression indicates that the prototype of manuscript **A** had a Chapter 9. Thus it appears that at least part of the Arabic *lacunae* are due to omissions of the final copyist, not of the prototype.

437 Latin Version: "*Court Attendant's Vessel*"—see note 273.

438 *The Latin version of this chapter is very different from the Arabic one, as shown by its translation in the following appendix. It also is more plausible, and is more similar to Hero's text, I, 7.* About the contents of Chapter 11, see Hammer-Jensen, pp. 480f. About the relation of Philo's clepsydra sieve to the archaic predecessors: Deonna, p. 218n, 219n.

439 The *qist* (Greek: *xestes*) has about 1 to 2 liters: Hinz[103], p. 50.

440 No doubt Philo meant: "... void of air as well as liquid." His explanation was quite correct, as he could presuppose terrestrial, atmospheric conditions.

Air cannot then enter the pipe, which is closed by the finger. The holes in the egg are closed by the liquid. The air cannot lift this liquid and enter into the egg, as it is lighter. The liquid cannot flow out, since its parts in the small openings are very small and have not enough weight to make them fall. Each hole is contained and isolated by the material of the egg. The egg is α, the pipe β, the holes γ. Here is the figure.[441]

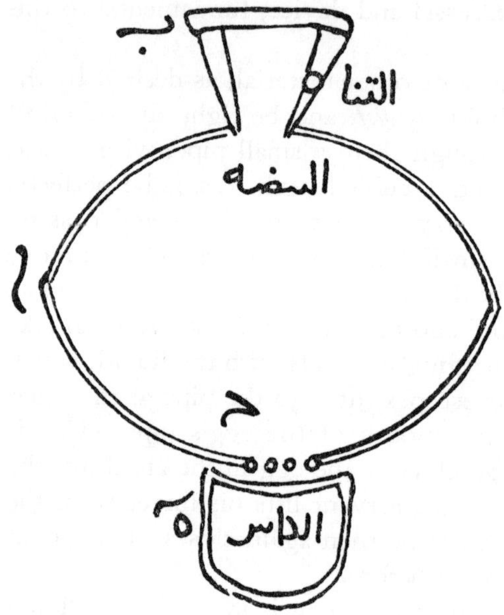

Figure 16. A. S. 3713, fo. 13b (middle). *Drawing of Chapter 11.* Legends and References:

ا (α): ["Egg"]
ب (β): ["Pipe", with "Mouth" at right]
ج (γ): ["Holes"]
ة (ε): "Container" to be filled

[Chapter 12. Other Clepsydras.]

The same things are proven by the following devices.

Pour water into a vase with narrow mouth, filling it up to the edge. Bring a leaf to the mouth of the vessel, hold it there with your hand, and overturn the vessel. The leaf will hang on, as if it were attached. You can withdraw your hand—no water flows out for a long time.

---

441 The drawing is clearly, and uniquely, derived from vessels of the type shown in Figure 2c. The Arabic copyist, in A. S. 3713, left the drawing unfinished, at the top; evidently he did not understand the design. The clepsydra drawings in the Latin copies are derived from miscellaneous versions of Figure 2a and b. They are quite varied in detail: they may show designs of clepsydras as used in the copyists' times (which then were called *hydrarpax*, *cantaplora*, Vestal's sieve, or even Aristotle's sieve).

You can also attach a sieve to the mouth of a [full] drinking-glass and then overturn it. No water issues.

[Latin Version:

11. *Another device should be mentioned. It provides a vessel "ab" with minute perforations "cb" in its bottom. After filling this vessel by pushing it into water, if you close the vessel's mouth with your finger [and raise the vessel], nothing comes out. Only when you remove your finger from its mouth does the water begin to discharge. Take care not to make the vessel too large, because the weight of the water, which descends with force, is too great.*

12. *Other demonstrations . . . Apply a leaf or the like to the mouth of a full vessel of this type and then invert the entire vessel; the leaf remains there for some time . . . Or if you attach a sieve to the mouth and then invert, exactly nothing comes out. The reason is that the water, running back and forth, fills the small perforations but does not descend further, since the air does not suffer it: thus the water does not leave. It has also been shown above that nothing can be void: therefore, while air has no place to enter, the water remains standing, not descending.*]

Editorial Note to Chapters 11 and 12

One of the *lacunae* of **A** begins in Chapter 9 and extends into Chapter 10; then comes another *lacuna*, as Chapters 11 and 12 do not follow at the end of Chapter 10. This end is followed directly by Chapters 13, 14, 15 and the beginning of Chapter 16, on authentic folios 8a to 12a.

Only in the middle of Chapter 16, the scribe of **A** inserts Chapters 11 and 12. He places the drawing of Chapter 11 in the middle of folio 13b. Thereafter, he resumes Chapter 16, in the middle of folio 14a, with a fragment of an Arabic chapter title or introduction, thereby marking the end of his displaced Chapter 12.

Carra de Vaux found and translated the text of Chapters 11, 12; he inserted it where it no doubt belongs, and where the present edition also inserts it; he made no remark about the confused sequence of chapters. Also, he evidently did not recognize the drawing on folio 13b as illustration of Chapter 11; otherwise he would not have reconstructed the drawing as showing a tube continuing the *long* axis of the "egg of copper or silver or other material."

Comparison with the Latin copies shows that they have a *lacuna* which begins substantially where the one of **A** ends: it is in no way affected by the specific defects of the Arabic tradition, in Chapters 4 to 12, but affects the descriptions of Chapters 13 to 16, in Latin copies **D, N** and **Z,** and affects both texts and drawings of these latter chapters in the other Latin copies. The probable explanation has been noted in the preceding notes dealing with the scope of the documents: the confusion seems to have begun as a result of poor preservation of the archetype; to have affected the Arabic prototype most seriously; and to have differently, independently affected the Latin prototype.

[Chapter 13. Liquid Dispenser with Siphon Outlet and Vent, Controlled by Pressure or Suction.][442]

Let us build another vessel of great elegance and beauty.

Let us take a vessel which has the form of a pitcher, or a copper jug, or any other desired form, with a long neck which has a single [hole, with a] cover. The bottom of the vessel is perforated by a number of small holes. The vessel has a spout similar to that of a water jug. The lower part of the vessel has a plane sheet soldered into it, separated from the bottom of the jug by the space of one inch. This sheet is perforated by a single hole. Provide a siphon pipe, shaped like a hook-knife for pruning trees.[443] Let it be two-thirds as high as the vessel [and adapted to discharge through the hole in the plane sheet].

Fill the vessel with water. Tightly close the hole in its upper part. If we then try to pour out water, nothing comes so long as the water does not reach the top level of the siphon. Only when it reaches that level, all of it disappears, entering the space below the siphon [and leaving through the holes.]

The water should, however, be kept at a level just short of the siphon's top. If we then want the water to issue from the bottom of this vessel, we must blow strongly into the spout of the vessel, applying both lips to it. The air then moves the water, which then has no [other] outlet; that is: the air pushes the water into a part of the siphon which is empty [of water] and full of air. The water then leaves by the small holes in the bottom of the

---

[442] An antecedent of Chapter 13 may be found in one of the archaic surprise vessels, Figure 2e. Hero has no direct equivalent; however, see his Chapter I, 24. Similar: Ibn Musa, Chapter 12. Most of the Latin copies omit this chapter, but **N** has its drawing and a brief description; **D** has the drawing only; and **W** has a fragment of the text, on fo. 94v: "Si [vis] ut restringatur [aqua] per decurlum [deorsum] fluere, subtus infla, ut autem refluat, suge." About the spout construction, see Chapter 23.

[443] A siphon pipe shaped like a "hook-knife" is also mentioned, for example, in Chapter 27.

vessel. If we then want to stop the water, we draw air with our mouth into our throat, through the spout of the vessel, sucking strongly. The air causes water to arrive in the spout and intercepts it from the siphon. The outflow then ceases. The pneumatic effect is admirable.

The vessel is marked α, its neck β, the hole therein γ, the cover for this hole δ, the spout of the device ε, the siphon tube $\zeta\eta$, the hole in the sheet ϑ, the sheet o, the small holes $\iota\varkappa$, the water level [for self-starting of the siphon] λ.

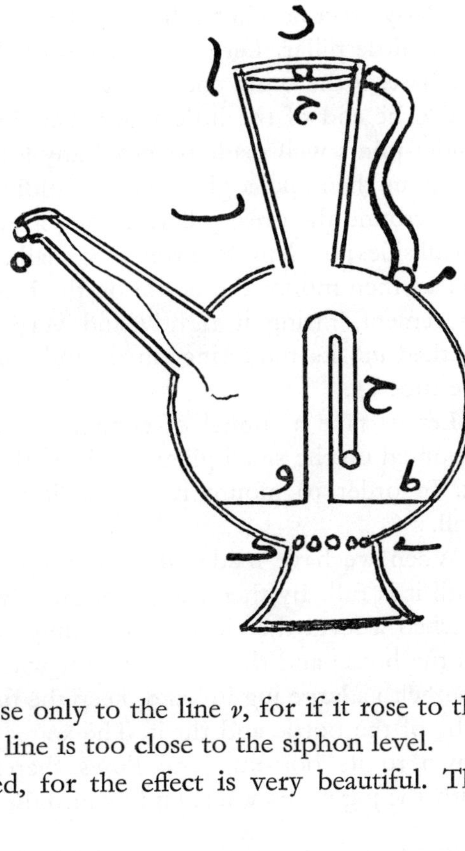

Figure 17. A. S. 3713, fo. 9a (top). *Drawing of Chapter 13* (formerly Chapter IX: note number 9 in upper right corner). Reference Letters:
- ا (α): ["Vessel"]
- ب (β): ["Neck"]
- ج (γ): ["Hole" in Cover]
- د (δ): ["Cover"]
- ه (ε): ["Spout"]
- ز ح ($\zeta\eta$): ["Siphon"]
- ط و (ϑo): ["Sheet", "Holes"]
- ك ع ($\iota\varkappa$): ["Small Holes"]

The water should be allowed to rise only to the line ν, for if it rose to the line λ, all of it would issue since this line is too close to the siphon level.

Let this be known and considered, for the effect is very beautiful. The illustration is on the other side.[444]

[Latin Version:

Second Figure (?) To force the water to flow down, blow into (the spout). To return it, suck (it out of the spout).]

---

[444] In **A,** the illustration of Chapter 13 is, indeed, overleaf.

[Chapter 14. Siphon Demonstrator with Outer Down-Pipe and Central Riser. Filled by a Funnel Unit.][445]

We also want to connect water to water, lifting one part by the other, as the other returns to the place of the one.

Let us prepare a vessel. Let there be a tube of brass, one and a half spans long, which we will call the base. Let us join to it a tube of lead, a cubit long, well adjusted, perpendicularly standing on the [one] end of the base, and which we will call the great pillar. On the other end of the base let us similarly erect another tube, of a length of four finger widths, and let us call it the little pillar. Let γ be the end of the great pillar, η its solder joint with the base, ϑ the other end of the base, with the solder joint to the little pillar, and ι the end of the little pillar. Let the pillars be tight all around and their solder-joints well made to avoid any leak.

Let us then take a glass bottle, holding one-half *qist*, and let us take a tube, to surround the great pillar, leaving around it a circular space one finger wide on all sides; the length of this tube shall correspond to that of the great pillar.[446] Let us then mount the bottle on the head of the tube by means of wax, pitch or cement, fitting it tightly and very carefully on all sides. The bottle is marked $\alpha\beta$, its mounting on the tube $\delta\varepsilon$, the water outlet at the bottom of the tube $\zeta\eta$.

Let us take a funnel of copper or other material, with its tube carefully mounted on the small pillar; let it be tight on all sides. This funnel is marked $o\lambda$. In order to mount it we dip it in wax and place it so that it adheres well.

When we have made all these preparations, we pour water in the funnel until it is full. By then the water already has entered into the great pillar and reached a level therein corresponding to that in the funnel. You must then fill the bottle and the tube below it with water. For this purpose have someone, with a large jug in hand, keep the funnel filled, while you rapidly take the tube of the bottle and tilt it. The water in it then surrounds the great pillar, down to its bottom, [and flows therefrom]. While you do this, the man with the jug pours water rapidly into the funnel. Water issues from the bottom

[445] The introductory sentence of this chapter as well as the end of its functional description sounds as if the author intended to lift water, by siphon action, into the glass bottle (after an initial starting of this action, by tilting the entire unit). The chapter, which is written in unusually inept terms, has no parallel in Hero's book, and probably is of a later time. Bottles of blown glass became known only about Hero's time: Rostovtzeff, pp. 1538f.

[446] The water flows from the open end of this tube: a messy operation.

end of the tube at ζ; what is in the funnel goes down [this tube], as the water in the bottle attracts what is in the great pillar, which leaves [this pillar], being connected with the first mentioned water. It continues to issue at ζ. Thus, even if one pours water into the funnel for a whole year, the flow [out of the tube] continues without interruption. All this can be seen [through the glass bottle].

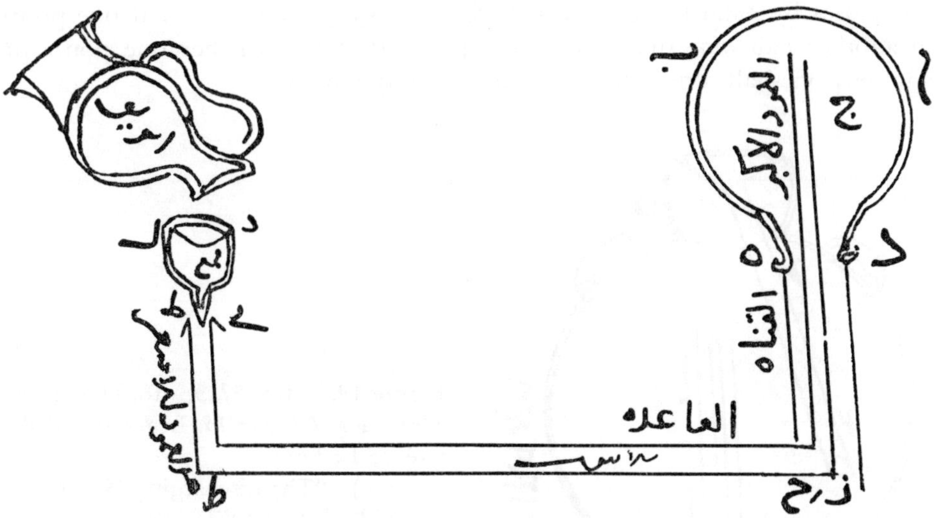

Figure 18. A. S. 3713, fo. 10b (top). *Drawing of Chapter 14.* Legends and Reference Letters:

From right top to bottom:
ا ب (αβ): "Glass Bottle"; "Sphere"
ج (γ): ["Great Pillar"]
د ه (δε): ["Mounting of Bottle"]
ز (ζ): ["Outlet"]
[horizontally]: "The Base"

From left bottom to top:
[horizontally]: "The Channel"
[vertically]: "The Little Pillar"
ع (ι): ["End of Little Pillar"]
و ل (ολ): ["Funnel"]
[in top vessel]: "The Pitcher"

Similarly, water adheres to air [and] air joins water to water. Let this be known and demonstrated.[447] You will find it admirable.

Here is the illustration [of the apparatus described in this chapter].

---

[447] It is demonstrated in Chapter 4b. At one time, that chapter may have followed at or after this point. Also see Chart 5.

[Chapter 15. Liquid Dispenser with Siphon Outlet and Air Vent Control.][448]

Let us prepare another vessel, wherein discharge occurs automatically. It has the form of a jug, but nothing prevents us from giving it another form. This vessel has a capacity of 6 *qist*. If one pours ½ *qist* into it [it seems that] the water remains and is not discharged, but if one adds to it, the water is discharged. [Actually], in the neck of this vessel a small hole is formed, which one closes by hand in order that discharge may not take place. If one pours in another half *qist*, the water flows [out]. It flows whether one pours [in another] one-half, one-third, one-quart or any amount.

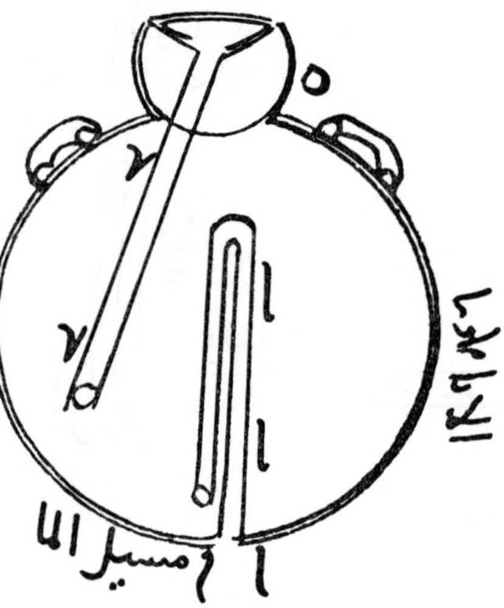

Figure 19. A. S. 3713, fo. 11b (top). *Drawing of Chapter 15.* Legends and Reference Letters:
ا ا (αα): "The discharging Water"
ب ب (ββ): ["Siphon"]
ج ج (γγ): ["Tube"]
ه (ε): ["Hole"]
At right: "The Container"

In the construction of this vessel, a hole is pierced in its lower part. A so-called intermittent siphon is mounted there. When this is done and the vessel is filled, the water discharges. It is [also] necessary to introduce into this vessel a tube, going from its opening [on top] to a point near the bottom, and being very tight. The discharging water is marked αα, the intermittent siphon ββ, the tube entering through the vessel's opening, and the hole in the neck of the vessel γγ, and the hole in the neck of the vessel ε. Now we have explained the construction.

If one pours ½ *qist* in this vessel, [the liquid rises only a small distance above the siphon inlet, and] nothing discharges through the open siphon,

---

448 Cf. Hero, Chapter I, 35. The drawing also occurs in **D.**

even when the hole is open. Air leaves through the hole. However, if one closes the hole with his finger, pours another ½ *qist* into the vessel and then leaves the hole open, the water discharges at once. The reason is this: while the hole is closed, the air can go nowhere but is compressed [and thereby presses water] into the intermittent siphon when more water is poured in, causing it to rise above its former level. When the finger then is removed from the hole, the [compressed] air returns to the place emptied thereof [and thus ceases to press water into the siphon, but the siphon has been started and continues to discharge]. The operation is elegant. As soon as one places one's finger on the hole [after starting the siphon], the water ceases flowing. If one reverses the vessel, one might believe it is void and nothing is inside. Here is the figure.

[Chapter 16. Demonstrating Horn Vessel with Glass Hat for Filling, Siphoning and Emptying Actions.][449]

Let us make a beautiful and enjoyable vessel. A horn is prepared as shown at $\alpha\beta\gamma\delta\varepsilon$. Above the receptacle of this vessel, a hat is placed, like that which the Egyptians place on the heads of their statues, similar to a long mitra, called ζ. Below this hat, at the open end of the horn—at $\alpha\beta$—, there is an inlet receptacle.[450] It is large enough to pour water into it, which then runs into the vessel.[451] The resulting effect is very strange. It is obtained by the vessel alone, which can be suspended from the top of the hat and need not be touched at all. The vessel is filled. When the water [rising in the horn] is about to reach the border of the horn, one sees it rise [also] from the bottom of the glass; it then stops and goes back, as it then flows out at the outlet, totally emptying the vessel. This is manifest by virtue of the glass and is visible to the eyes.

In the construction of the unit, container $\alpha\beta\gamma\delta\varepsilon$ and hat ζ are used. The bottom of the hat is connected with the horn, with accuracy, by the [pipe] parts $\eta\vartheta$, [at] a closure of the horn shown by the straight line $\gamma\delta$. Two copper

---

449 Carra de Vaux saw rightly that there is some similarity between this apparatus and the Heronian horn vessel II, 13 (a variant of his I, 19 and II, 28), but showed an erroneous, inoperable reconstruction. Beck recognized its inoperability, but assumed that Philo's vessel looked approximately like the vessel shown by Carra de Vaux.

450 Carra de Vaux called it "an orifice." This interpretation disagrees with the authentic drawing, and is not required by the Arabic text.

451 Through a central orifice, without reference letter, as correctly stated by Carra de Vaux.

pipes are used. One is connected to the bottom of the hat at ϑ and extended down through the closure $\gamma\delta$; its end passes through the horn and beyond, emerging from this vessel at ε.

The closure adheres exactly at $\gamma\delta$. The sidewalls of the horn are so arranged that, when water is poured in, it does not flow out directly, from the inside,

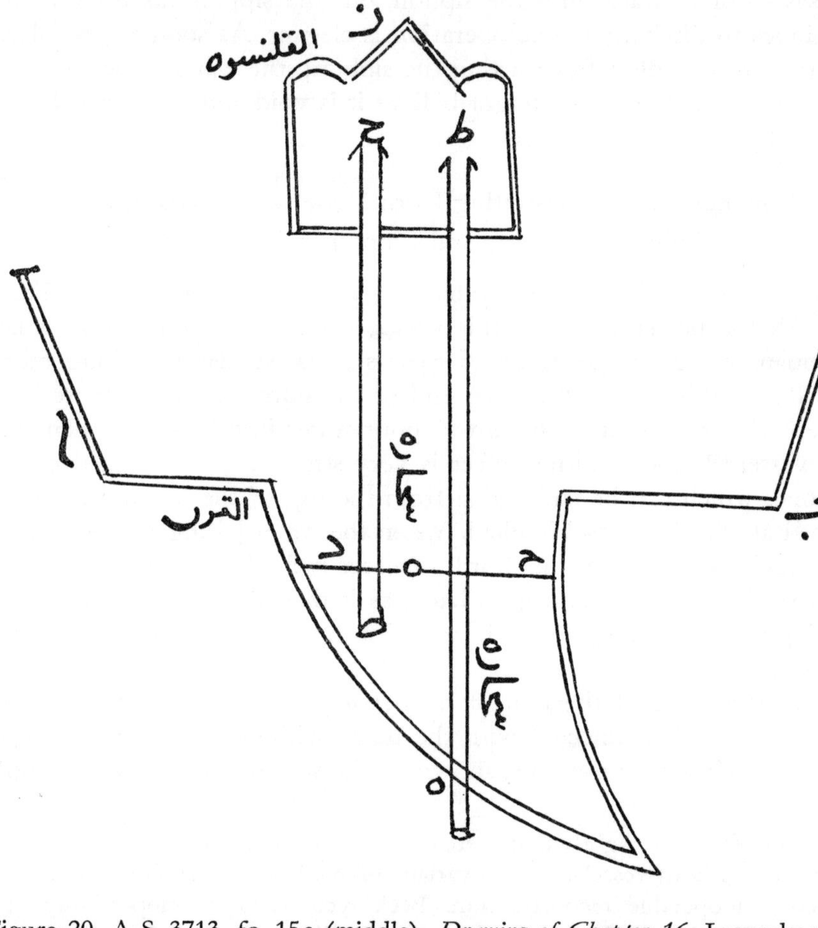

Figure 20. A.S. 3713, fo. 15a (middle). *Drawing of Chapter 16.* Legends and Reference Letters, from upper left to lower center:

ز (ζ): "The Hat"

ا ب ($\alpha\beta$): [Pan, attached to the wide end of the Horn]

د ه ج ($\delta\varepsilon\gamma$): "The Horn"

ط ه ($\vartheta\varepsilon$): [Pipe "parts $\eta\vartheta$"?]

that is, from the pointed end.[452] The second pipe is connected to the bottom of the hat, at η, and adheres tightly. It goes down somewhat below the closure ($\gamma\delta$), in order that it may receive water [from the horn]. The end of the longer pipe, which goes to the outside and is marked ε, should be somewhat more restricted than the other end.[453]

When the vessel has been prepared, as we have now described, water or wine is poured on the receptable [α]β. The compartment below the closure then fills. As it fills, the second pipe fills to a level near ϑ.[454] This pipe then emits a water fountain, springing forth [in the hat, due to compression of air in the upper part of the horn]. The water from this fountain rises up [in the hat] until the outlet pipe fills, and finally flows out through the same. See the figure.

That the water remains [in the hat], until the outlet pipe fills, and that it does so outside this pipe, is because this pipe is longer [in the hat as well as below it] than the one that goes from η to ι.[455] After the filling of the outlet pipe, [siphon action occurs from the short pipe through the hat to the outlet pipe, wherein] the water moves violently and entrains part of the air in the hat. The air remaining there is rarified. A greater [than the normal, distributed] amount of void is introduced. As the water meets the very rarified air, its rising is promoted.[456] This becomes evident if you consider it; give it thought.

[Chapter 17. Liquid Dispenser with Flooded Vent Control, for Maintenance of a Constant Level.][457]

In order to establish the truth of our doctrine by many examples and to explain the construction of the devices, we will show the following. Prepare

---

[452] Carra de Vaux: "... does not flow out at all from the ... end." This translation is not required by the Arabic text.

[453] Probably to slow up the flow and facilitate the observation.

[454] This level is above the horn unit. The arrangement came to be called "Hero's Fountain," in the Italian Renaissance.

[455] As to this point of the Arabic and Latin traditions (shown by **A** and **D**), the drawings are not quite correct according to the text. It is also incorrect when Carra de Vaux says that water remains "in" the outlet pipe.

[456] This is an imperfect description of the siphon effect. However, it shows good recognition of the fact that a partial vacuum exists at the turning point of the siphon.

[457] A greatly modified variant of Chapter 17 is shown by Hero, I, 19. Even greater modifications are described by Ibn Musa, with illustrations (Wiedemann-Hauser, Ibn Musa Drinking Vessels) wherein the flooded vent disappears, pp. 287–291, and float valves appear instead, pp. 270–286. Drachmann, pp. 51–54, unaware of the authentic drawings in **A, C** and **Z** and of the drawings of Ibn Musa, would reconstruct the Philonian device in a form similar to the Heronian one.

an elongated wooden base. On one side thereof place a long, rounded wood piece. On [one end of] the base, place a jug. On one side of the jug provide [a horizontal liquid outlet pipe, and inside the jug, as a vent] a pipe directed downwards. On the other end of the base place a cup. When you fill the jug with wine [while] the [vent] pipe is open, some of the wine flows and fills the cup. Then it closes [the vent] and nothing flows any more, until one takes some of the wine. When one takes it, it again flows, in equal amount. This repeats itself, as described, until the jug has no wine any more. The device is very elegant.

Its construction is as follows. The long, visible base is marked *αα*, the long round wood *ββ*, the jug *γγ*, the cup below the pipe *δδ*. Below the cup there is needed another vessel, of glass or wood, similar to the round wood on the base, to support the cup. This other round wood must be hollow, as that described before.[458] It must be somewhat higher than that one, to let a passage enter its side from within.[459] For proper operation this vessel engages and holds [the upper edge of] the cup from below.[460] The vessel provided by this round wood is marked *ε*. The cup rests on it at *δ*. The [vent] conduit through both [woods] is *ϑ*, the jug's outlet pipe, *ϰ*.

The jug must have a pipe entering its bottom, [as noted], and this pipe must have one end rising to the neck of the jug. Its other end must go through the rounded wood and the wood vessel connected with it.[461] The mouth of the jug must have a cover, which must be tight. This pipe is *ζ*. It must be made as described. Then the jug can be placed above the [end of the] long round wood. The upper end of the pipe is reversed downwards [to avoid filling it with liquid when one fills the jug].

---

458 It has *not* been so described, and is not so illustrated with any clarity, in the Arabic or Latin Philo. The archeological findings (Deonna, pp. 221–223) show that pedestals were developed, in classic and pre-classic times, for example for the vase shown here in Figure 2c, in such forms as 2d, f. Philo and Hero then developed various pedestals, one of which is shown here in Chapter 27, generally for purposes of concealment and surprise. Probably the present pedestal served to allow concealment of the vent pipe.

459 Probably in order to avoid formation of a liquid-retaining U-pipe, in the vent, which makes the venting process impossible. As rightly stated by Hammer-Jensen, pp. 485–487, Carra de Vaux overlooked this implication of the text. The authentic drawing confirms it. Beck was misled by the Carra de Vaux edition, in this chapter and the preceding one.

460 Literally: "For this purpose the vessel must adhere to the cup from below." The present interpretation is suggested by the authentic drawing.

461 Literally: ". . . the vessel supported on it."

A spout is placed on the mouth of the [liquid] outlet pipe. Wine is poured in the jug. When it is full, one may close the spout with one's hand, [but] one [also may] open it if the vent pipe in the wood pieces is ready for it.[462] The liquid then runs into the cup. When the cup is full, the rounded wood

Figure 21. A. S. 3713, fo. 17a (middle). *Drawing of Chapter 17.* Reference Letters and Legends:

ا ا ($\alpha\alpha$): ["Base"]
ب ب ($\beta\beta$): ["Round Wood"]
ج ج ($\gamma\gamma$): "The Jar"
د د ($\delta\delta$): ["Cup"]
ز (ζ): [Vent Pipe]
ك ($\varkappa$): [Reverse Bend]
ط (ϑ): "The Wine Outlet"

vessel also is full, through a hole in the bottom of the cup. When the wine has risen to the mouth of the cup's pipe [-vent], the outflow ceases, as no air can thereafter enter the jug any more. When one then takes wine from the cup,

[462] This translation attempts to make the Arabic text as plausible as it can be made. Literally, as given by Carra de Vaux, it seems unintelligible.

the vent pipe opens.[463] Wine then flows again. This repeats itself until the jug is empty. Here is the figure.

[Latin Version:[464]

Let there be made another vessel "ab" with a narrow mouth and let it be apertured in its bottom at "d". Here let a pipe enter vertically, with a return bend close to the neck of the vessel; this pipe shall be "cde". Let its part "de" extend upright in the vessel and let its other part "cd" extend outside. Let its far end "c" enter another vessel "ghz", which is perforated below. Let these two vessels be on a hollow pedestal "mn", so that water which issues from the bottom of the vessel "ghz", at "t", may enter there and nowhere else. Let still another pipe "pqr" be adapted to the vessel "ab" to issue at point "r" and extend so that its end "p" is located above vessel "ghz".

If you fill vessel "ab" with water, up to the top of "de", and close its mouth with a cover "xy", water will continuously flow into vessel "ghz" until the pedestal on which it stands is full and the water rises to "c". When this point is full of water, air has no way any more to enter vessel "ab"; therefore water ceases to run at "p", until it is removed from "c", whereon it issues as before. Thus it will always stand at the same level].

[Chapter 18. Constant Level Liquid Dispenser with Flooded Vent Control and Column Supports.][465]

Construction of another analogous vessel, of a design simpler to understand.[466] Let us take a vessel structure wherein the largest vessel, $\alpha\beta$, serves

[463] For this purpose the cup must be somewhat deeper than the Arabic drawing shows. The expression "cup" has been retained here, but "vessel" or "receiver" may be better. It does not seem that Philo expected to remove the "cup" from its mounting, to empty it.

[464] The Latin version of **N** is translated here. It coincides in substance with the other versions of traditions L–I, L–II, and is almost as unclear as the Arabic one. It differs in that it states, with grammatic clarity (in spite of remaining, technical vagueness), that wine runs from a lower perforation of the cup (ghz) into the "hollow pedestal" (mn). *Evidently the translators were differently confused by corrupt archetype expressions.* Some western copyists (**C, I, Y**) seemed to recognize the confusion about the perforated cup and the hollow pedestal: they reinterpret or omit the pedestal.

[465] A variant of Chapter 17, replacing the pipe-concealing pedestal by a column. Chapter 19 replaces it by a wall.

[466] Simpler or lighter? It seems that the archetype used the ambiguous word *elaphrios*, which the Latin translator rendered as *levius*, "lighter of weight," while the Arabic translator rendered it as "simpler . . . for those who wish to study this art." The Arabic variant seems improbable; Philo expressly wrote his entire work for

as reservoir for the water and rests on three or four columns.[467] In one of the front columns, or in any other column if desired, is an entirely hidden pipe ϑ extending through the hollow of the column and into the interior of the vessel, up to its top. Take a cup $\iota\varkappa$, tied to the columns and fixed to them. Make a copper lion π, disposed above the cup.[468] Place another pipe in the hollow of another column, extending into the hollow of tbe lion and going to his mouth. The water is marked ζ, the lion's mouth and water outlet η. On top of the reservoir place a well-made, tight plug δ, free of air leakage.

When this has been prepared, we close the lion's mouth with a finger. We open top δ of the reservoir and fill the reservoir with wine up to mark ε. Then we replace the reservoir cover and seal it very carefully and solidly. This is to be noted. Here is the figure.

When we remove our hand from the lion's mouth, wine comes from the lion's mouth and into the cup. It stops only when it covers hole ϑ. When this hole is covered, the outflow of wine from the lion's mouth ceases, the lion does not dispense any more, and nothing flows, until enough has been taken from the cup to uncover hole ϑ. Thereupon the lion spouts liquid in quantities equal to the material removed. This continues until nothing remains in the reservoir. The amount needed to uncover hole ϑ depends on the operator's preference: 2 *ritl* or 3, or less or more.[469]

This device is the clearest, simplest, most elegant and most easy to operate, for those who wish to study this art.

"learned men" interested in ideas of "philosophers", not for a school of model builders or demonstrators, where some models were handled by beginners (commençants: Carra de Vaux). The model of Chapter 18 is not inherently lighter than that of Chapter 17, nor is it any easier to make or explain. However, both are models, lighter than the outdoor "wall" construction of Chapter 19. It seems that Philo described Chapters 19, 18, 17, in this sequence; that the archetype reversed the sequence and confused Chapter 17; that both Latin and Arabic interpreters were somewhat misled; but that *the Latin "levius" is better than the Arabic "simpler"*.

467 The Arabic drawing shows these columns in a pseudo-Corinthian form, similar to the form shown in Figure 1. As noted above, this form may be traced, perhaps, to lost illustration of pre-Philonian books. It certainly occurs also in Carolingian art.

468 Carra de Vaux designated the lion as λ. He may have overlooked the Arabic π, between the lion's eyes, and may have misinterpreted an Arabic α, between the lion's head and the top container, as an Arabic λ.

469 The *ritl* (Greek *litron*) is c. one kilogram: Hinz[103], pp. 27–33.

Figure 22. A. S. 3713, fo. 18a (full page, except one line above). *Drawing of Chapter 18.* Reference Letters and Legends:

ا (α): ["Largest Vessel"]
Marked 3 times

د (δ): "The Plug"

ط (ϑ): "The [hidden] Pipe."
Marked 3 times.

ز (ζ): "The Water"

ع ك ($\iota\varkappa$): "Receiver of the Water"

پ : ["Lion." Marked in the Lion's Forehead]

ح : ["Lion's Mouth"]

[Latin Version:

Other Example. We can make another vessel "a . . ." of the same series, but of lighter construction, on two columns "d" and "e". Let its mouth be at "c", and the tube which issues, ". . . p". Let one of the columns be hollow. Through it, let a tube enter into vessel "ab", which extends to vessel "ghz", that is, tube "qfz." Let vessel "ghz" be positioned below the mouth of the tube. Let the tube be fitted at "p", so that the water issuing may enter . . . So, when vessel "ab" has been filled to level "c" and its mouth "c" has been sealed, the water will issue through "p" until the vessel "ghz" again is filled. It will be entirely the same process in this vessel as in the preceding one.]

[Chapter 19. Constant Level Dispenser with Flooded Vent and Concealing Wall.]

Construction of another vessel of this type. It is built in such size as the designer wishes. The vessel is filled with wine. At its top is a place of connection to the vessel [for venting a supply container]. When the servant takes some liquid from the [other] vessel, into which the liquid has beeen poured, an amount equal to that taken out comes back into this vessel. This repeats itself until the company has consumed the liquid.

The construction is as follows. A container is so placed, behind a wall, that the cup-bearer can reach it [to fill it]. One makes a pipe fitted to the bottom of the container, with an upwardly facing end. The pipe goes through the wall. Where it opens into the container, it has a sieve. The vessel below this container—the one which receives the transferred wine—must have a small hole near its edge. A copper tube, soldered to this hole, goes through the wall and into the container, almost to its mouth.

One pours wine into the container, in the desired amount, almost up to the mouth. One then closes this mouth all around to prevent leakage of air. The lower vessel fills, as one leaves the container's outlet open. When a drinker has taken some wine from the [lower] vessel, there again enters the same amount as the one taken; then it stops, until someone drinks again.

Note that the flow stops, when the vessel is full, because nothing can flow from the end of the tube which brings the wine, so long as it[470] receives no air. When one has taken enough to expose the tube, so much enters again as has been taken, since the air then reaches the tube. This is repeated any number of times, as we have explained it.

---

470 The large container.

The large container is marked α, the upwardly facing pipe β, the receiver of wine γ, the tube $\delta\varepsilon$, the pipe from the receiver to the board of the vessel $\zeta\eta$,[471] the wine level ϑ, and the cover ι.

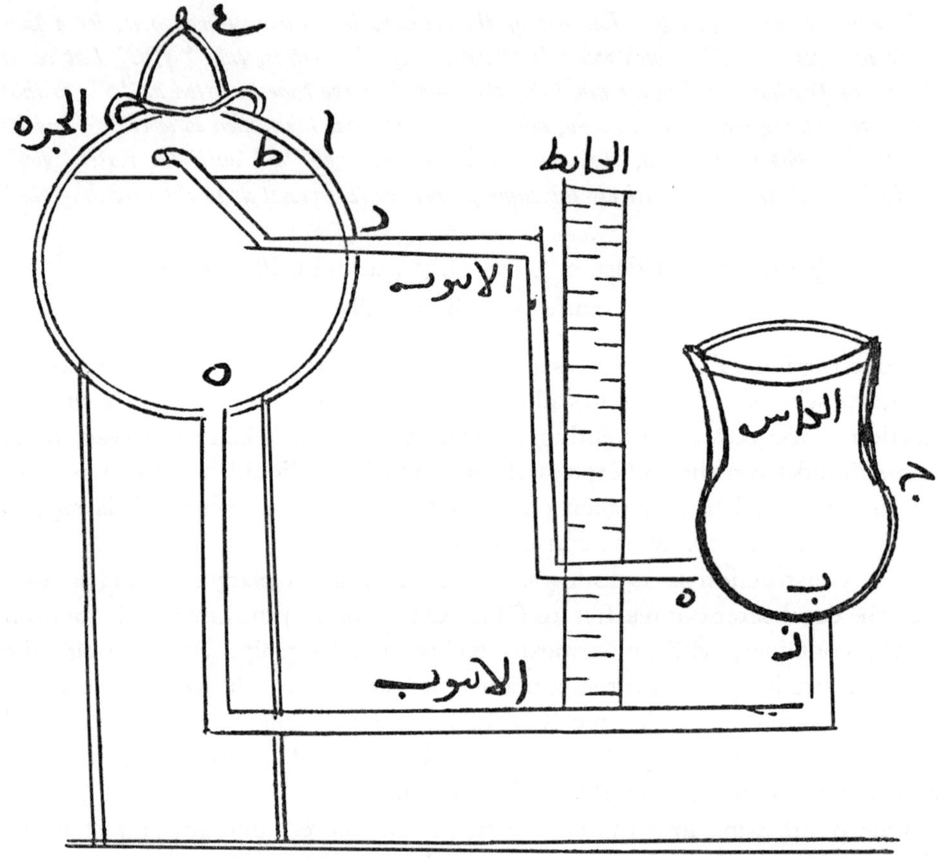

Figure 23. A. S. 3713, fo. 20a (middle). *Drawing of Chapter 19.* Legends and Reference Letters:

ا (α): "The Container"
ز ه , ب ه ($\beta\varepsilon$, $\zeta\varepsilon$): "The Pipe"
د ه ($\delta\varepsilon$): "The Tube"
ج (γ): "The Drinking Glass" ["Receiver"]
ط (ϑ): ["Wine Level"]
ع (ι): "The Cover"
Upper Middle: "The Wall"
Lower Middle: "The Base"

---

471 Three errors lie in the expression: "The pipe from the receiver to the board of the vessel [is marked] $\zeta\eta$." (1) This pipe has just been defined as "the tube $\delta\varepsilon$;" thus it is confusing to also define it as $\zeta\eta$. (2) The vessel which has a pipe or tube joined

The three devices that we have described are constant level devices. You can use them as you desire, [also] for baths, washstands, and lamps. All of them work in the same way.

[Latin Version:

Other Example. There are made other vessels "ab", "ghz" with their tubes "p" or "cde", in the same way as before, except that partitions "st" are interposed between them. When the vessel "ab" has been filled with water almost to "p" and when its mouth "o" has been sealed, the contents will descend through "cd". As it, from there on, enters into vessel "gh" until it reaches "r", everything will be as before.]

[Chapter 20. Constant Level Oil Dispenser with Flooded Vent Control, for Lamps][472]

Construction of a Constant Level Device for a beautiful lamp. Take a vessel formed like a pine cone, holding the desired amount of oil. Provide a pipe in it and extending from it. Let it be like a testing-tube.[473] Make a cluster of two, three or four lamps. In its middle provide an upright pipe, with a hole entering the side of one of the lamps. The test tube is placed into this pipe.[474]. The oil reservoir has an outlet tube to each lamp, in form of a trunk [of an elephant].

Fill the pine cone with oil through the test tube. Mount the fixed center-tube of the lamp cluster on it.[475] The trunks fill the lamps, until the oil has

---

to its board has just been defined as the receiver of wine; thus it is confusing to speak of a pipe from it to the receiver. (3) The drawing does not show the point η, mentioned here, at all, and shows ζ only by confusion with δ, which is not at the board. *The Latin translation is free of these multiple errors.*

[472] Basically about this chapter: Hammer-Jensen, p. 494; also see Drachmann, pp. 122–125. About the archaeology of oil-burning lamps: J. Toutain in Daremberg-Saglio, *s.v.* lucerna, lychnus: central filling fixtures, and pine-cones, were known before Philo. *The Latin and Arabic traditions differ*, in that the former shows a lamp wherein oil reaches the burning point by capillary action (as does Hero, II, 22–24), while the Arabic drawing shows an oil level hydrostatically maintained at the burning point. Ibn Musa shows both arrangements: drawings 95, 96. A very vague disclosure of the device is given by Cardanus[86], p. 5.

[473] Probably meaning, a testing-tube for wine.

[474] Probably meaning: it has sliding but air-tight fit with the upper end portion of this pipe, above the hole.

[475] That is: slide this test-tube into the fixed center pipe of the lamp cluster, establishing an air-tight joint between the tube and the pipe.

risen to the hole and closed it; then the flow from the trunks ceases. When the flames have consumed the oil, and the hole opens, the trunks discharge again. These actions continue until the oil is exhausted.

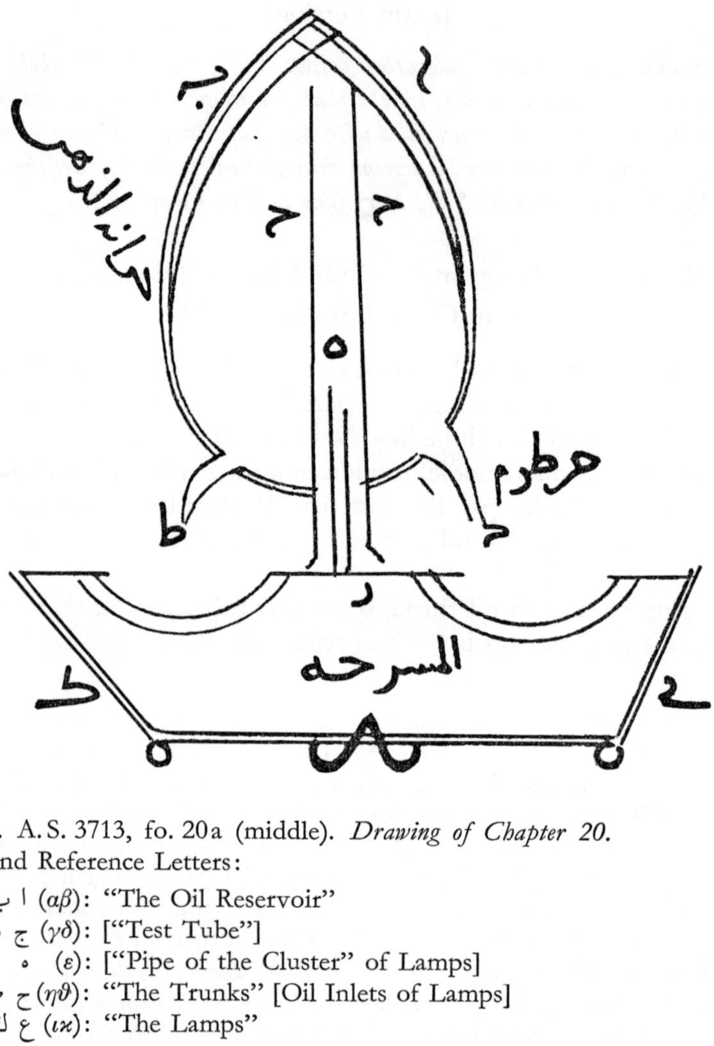

Figure 24. A. S. 3713, fo. 20a (middle). *Drawing of Chapter 20.*
Legends and Reference Letters:

ا ب ($\alpha\beta$): "The Oil Reservoir"
ج د ($\gamma\delta$): ["Test Tube"]
ه (ε): ["Pipe of the Cluster" of Lamps]
ح ط ($\eta\vartheta$): "The Trunks" [Oil Inlets of Lamps]
ي ك ($\iota\varkappa$): "The Lamps"

The reservoir is marked $\alpha\beta$, the test-tube $\gamma\delta$, the pipe of the cluster ε, its hole ζ, the trunks $\eta\vartheta$, and the Lamps $\iota\varkappa$.

There are many variations of the Constant Level Devices. You can modify them as you wish. Here is the figure.

[Latin Version:

Similarly, let there be made a vessel "abc", bounded by a single surface, with two tubes "cd", "be" at the sides. Let it also have a vertical tube inside, descending into a vessel "ghz". Let each vessel stand firmly in its place "l, m" on this tube, klmn. Let small parts of vessel "ghz" extend from the sides in form of crucibles, and be set below tubes "cd", "be"; let these small parts be "gt", "zs". When you then fill vessel "abc" up to a level "n", the liquid that you put in it will issue, both from "dc" into "sz" and through "be" into "gt", going at both places into the vessel ["ghz" until it reaches tube end] "ck". When it fills this end, the outflow at "de" stops.

Let the liquid in vessel "abc" be oil. Let a wick or [strip of] papyrus be fitted in vessel "ghz", where the oil is. As the oil in "ghz" is consumed by burning, it gradually descends from "abc" through "d", "e". These four instruments are of the same kind. This is the end.]

[Chapter 21. Multiple Drink Dispenser with Clepsydra Type Vent Controls][476]

Construction of another vessel. Its form is that of a carafe. Its body is made of clay from the island called Chios in Greek.[477] There is a faucet at the bottom of this carafe, for tapping off the contents.[478] Four kinds of liquid are poured into the carafe, and through this faucet, each is taken off separately, unmixed with the others. A first one is chosen by the servant; others can then be selected by the master.

The vessel is constructed as follows. Inside the Carafe, two partitions are disposed, which go from the bottom to the neck, one intersecting the other so that they together form four compartments. In the lower part of each compart-

[476] Cf. Hero I, 8, 23.

[477] Half-way between Byzantium and Rhodes. Not particularly known for its clay during the middle ages, when the island was Byzantine, but possibly known to Philo for some ceramic specialties. Stylistically, the vases of Chios had some special features.

[478] *The Latin version, below, speaks only of a "lower tube" and implies that it is controlled by clepsydra action, not by a valve.* The jar with such a faucet tube is attributed to Philo, by Drachmann, p. 67, 69, whereas the "ground cock" of Chapter 22 is said to be "new" and "Arabic": Hammer-Jensen, p. 426; Drachmann, p. 50. However, consideration must also be given to the ground pipe or piston construction of Chapter 20, the air-tight closures of Chapters 4 and 17–19, and the plug valves of Chapters 22 ff., 43 ff., as well as the more developed plug valves of Chapters 51 ff. The first such construction is explicitly, authentically attributed to Ctesibius; the last of them appear to belong to the early middle ages; others may belong to various intermediate times. Also see the Index, *s. v. smerismation.*

ment there is a partition, pierced by a hole which leads to [an undivided bottom chamber having] the outlet pipe, marked ε, from the bottom of the vessel to the outside. Let the [first] two partitions in their upper part, near the neck, be held together. Let a screen be formed.[479] Let it extend over all compartments, [Provide] in each [compartment] a single hole to the outside, on the circumference of the carafe, in the upper part of its belly.

One takes the vessel in hand, closing the four holes on the carafe's belly. The lower outlet is held closed by the servant's hand. When the compartments are filled with liquids,[480] and when the servant wants to pour out one of the liquids—wine, vinegar, milk or water—he lifts his finger from the hole

Figure 25. A. S. 3713, fo. 22b (top). *Drawing of Chapter 21.* Reference Letters:

ا ب ($\alpha\beta$): ["Carafe"]
ج د د ج ($\gamma\delta\delta\gamma$): ["Partitions"]
ع (η): ["Head"]
ه (ε): ["Outlet"]
ط ع ($\vartheta\iota$): ["Compartments"(?)]
ز م ($\zeta\mu$): ["Holes"(?)]

of the compartment in question, while leaving his fingers on the other holes. These admit no air; no liquid flows except the one selected. Later he can close that hole, open another, and produce flow from that compartment. He can continue this so long as he pleases.

Let the carafe be marked α, the partitions $\gamma\gamma$, $\delta\delta$. There are holes in the neck of the carafe, one for each compartment.[481] After admission of the liquids

[479] More correctly: a plate with four holes, one to each compartment(?)

[480] That is: each with one of the liquids.

[481] Carra de Vaux interprets that the head "turns so that a single hole in its bottom . . . successively coincides with each of the four holes."

through the carafe's head, one closes this.[482] The carafe's head is η. Two holes ζ are for two compartments ϑ, two μ for ι. The holes which the servant closes and opens with his fingers are the four holes v. This is what we wanted to explain. Here is the figure.

[Latin Version:

We shall make a vessel with lower aperture and lower tube "g", and entirely divided into two or four or more parts, by walls "a, b, cc, d, e." Let each of these [parts] be perforated, above and below. Let all be filled, with various liquors. If you close the upper perforations, nothing will issue below. If you open any one, above, that liquor will descend downwards, and issue through the tube "g" of the vessel.]

[Chapter 22. Multiple Drink Dispenser with Concealing Top and Rotary Outlet Valve][483]

Construction of another, more beautifully functioning vessel, again for dispensing four liquids. Let there be a vessel with four compartments, above a receptacle called the take-off. In the middle of the vessel is a pine-cone, which projects. In order to open all compartments, we turn this pine-cone part of one revolution; to close them, we turn it back. To open any two, or any one, or any three, we make further partial turns.

The vessel must be divided into the four compartments below a circular closure $\iota\kappa\lambda\mu$.[484]

In the middle is a cylinder, which forms the female part of the faucet valve. In its hollow, one places another cylinder, pierced by holes in four positions predetermined to be opposite a hole from one compartment. These are $\alpha\beta\gamma\delta$. This other cylinder forms the male part and is shown here. The female part is traversed by passages from the compartments, each passage coming from one of the compartments of the vessel. This female part is firmly mounted. In the bottom of the vessel is a small space or receiver into which the liquid outflow holes of the four containers discharge.

We take the other cylinder, called male, which turns in the female and is pierced by four holes $\alpha\beta\gamma\delta$ exactly corresponding to those of the female part.

---

[482] By the hand of a servant? Evidently the Arabic text, as well as the drawing, is very unclear in a technical sense.

[483] This is a variant of Chapter 21, which is also illustrated, although confusedly, in the Latin manuscript **D.** It also occurs in Hero, I, 23, and AlJazari, ed. Wiedemann-Hauser, Drinking Vessels, pp. 77–81.

[484] Odd reference characters are used for this device, in **A.** It seems from the record in **A** and **D** that this part of the archetype was confused.

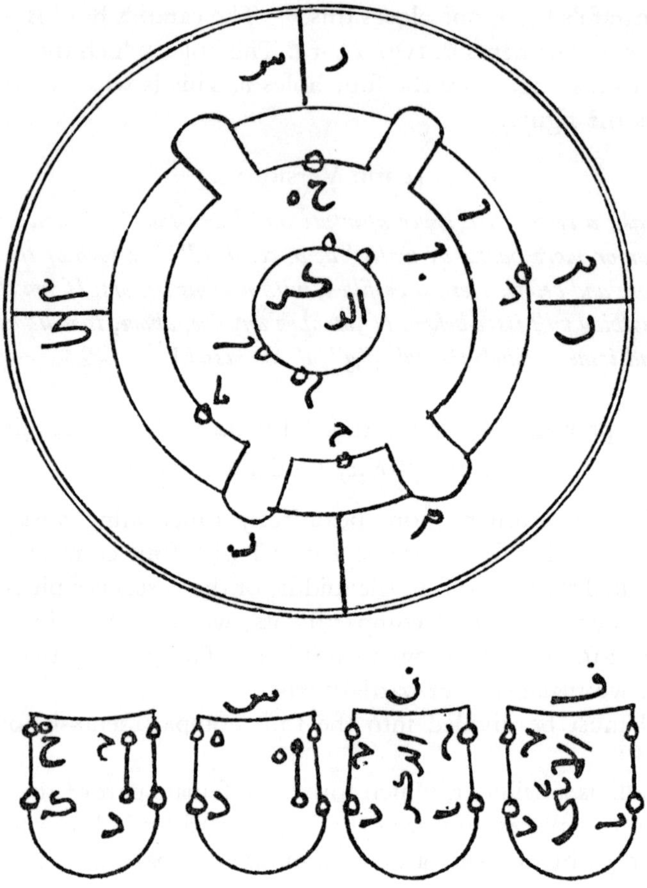

Figure 26. A. S. 3713, fo. 24b (full page). *Drawings of Chapter 22.* Reference Letters, in upper figure, outermost circle:

ع ك ل م (*ικλμ*): [“Circular Closure”(?)]
م ن ر س (*μνϱσ*): [?]
in upper figure, innermost circle:
ك ر (*κϱ*): [?]
in upper figure, intermediate circles, and lower figures:
ا ب ج د (*αβγδ*): [“Holes”(?)]
ز ل س (*ζλσ*): [?]

When you have these holes in front of $\alpha\beta\gamma\delta$, all liquids in the compartments are discharged, through the holes piercing the male part, including ε and ζ; but if you wish, hole α can be registered and opened, the other holes remaining closed. You can then turn [to] hole ε to close it off, or to δ; or ζ can be opened and the others closed. You can then re-register hole γ after closing off ζ: hole γ can open. You can, while leaving it closed, pierce hole η opposite γ, and ϑ opposite δ, to register when all others are closed.[485] To open all holes, leave α of the male part opposite α of the female; all are then open.

Under the male part, before mounting it, a collar can be mounted in the female part, to allow not only the turning and registering of the male part but also removal, if desired. At the head of the male part, place a square pine-cone, in the middle of the vessel, to turn it for opening or closing.

On top of all reservoirs place a screen, so that one may not know the inside works. Pour different liquids, to be dispensed at will through the holes: singly or together. Let marks be made on the vessel and pine-cone, so that we can register the marks to dispense something from the vessel: it will arrive as desired. Here is the figure.

[Chapter 23. Multiple Drink Dispenser with Concealed Rotary Inlet Valve][486]

Construction of a two-liquid-ewer. This is another vessel which dispenses two liquids, or three or four if preferred. It is easier to understand than the first.[487]

Take a well-made ewer of silver, tin or copper, with a long and strong neck. Divide it longitudinally into two halves by a partition of the same material, which continues along the neck to near the mouth. Here the neck is closed transversely. The closure of the neck has two holes $\alpha\beta$ pierced in it, each going to one half of the ewer.

The ewer is provided with a head, which turns on the rim of the neck and is carefully fitted to it, for proper operation of the ewer; it is longer than the neck

[485] The construction presupposes an accuracy of metal grinding far beyond that implied in Chapter 21 (note 478). Whether such accuracy was reached in ancient Egypt is an open question: Rostovtzeff, pp. 374ff.

[486] Deonna, Fig. 12 (present Fig. 2d) shows a partly hypothetically, reconstructed VI. century vase of similar type, signed by the potter Kolchos. Hero, I, 22 shows a similar vase, but omits the spout or spouts. Drachmann, p. 62, unaware of the Greek artifacts, considered the diagonally rising pipe-spout characteristic for the Arabic *ibriq* (which it is), and newly introduced in Arabic times (which it is not; Figure 2f.).

[487] What is meant with "the first" is unclear.

and is hollow. This head is closed at its bottom by a flat sheet. The sheet is pierced by a hole γ, which [in each position of the head] leads to only one-half of the ewer. At its top, the head has a screen, through which we can pour what we want; one cannot distinguish where it goes.

The ewer has an outlet pipe, which we will call *bulbullah* [Mrs. Nightingale].[488] It is of medium width and of careful construction. There is a second outlet pipe, which is longer and thinner. At the point where this latter pipe enters the first, it is only half as wide as that. The wide pipe is δ, the thin one, ε. The thin one is mounted on one of the halves of the ewer: on the inner half, ζ, near the handle. Here the partition is pierced. The thin outlet pipe is mounted there, and it then goes to the outside of the ewer, through the wide outlet pipe. That pipe is mounted on the other half, η, of the ewer. The thin pipe ends short of the wide one, concealing itself from any one who inspects the ewer and thereby adding to the effect of the demonstration. Thus we dispose two outlet pipes in one.

The ewer has a handle, wherein the principle of the operation resides. This handle resembles the two outlet pipes, in that it has one duct entering ewer half ζ and another entering η. Each of these ducts enters through the head and neck unit.[489] The two ducts are soldered together to form the handle while concealing the tubes. The tube for η has a hole ι; the tube for ζ has a hole $\varkappa$.

This completes the construction.

In order to pour water and wine into the ewer, we turn the head, which operates as a valve. We observe marks, formed on the ewer's neck and valve, to know that the hole in the bottom of the head overlies one or the other ewer division. Then we pour the desired liquid.

We then turn the head the other way, and the hole then overlies the other division. Again we pour whatever is desired. Another mark is formed at λ: we turn the valve to register it with a mark μ on the neck. Then the hole of the head is closed, whereby operation becomes possible.

Accordingly, to pour out the contents of the ewer, the servant takes the handle in his fist, as is usual with such vessels. To pour wine, he opens one of the two holes, as instructed, under his thumb. The wine then comes from the

---

[488] As shown by Wiedemann, ed. AlJazari, Drinking Vessels, p. 59, the word *bulbullah* was also applied to an open, channel-shaped spout. At the present point, a translator may have felt that the outlet pipe shown here is a *bulbullah* of a new style or construction.

[489] Specifically: through the neck, below the head, as shown. The drawing of Carra de Vaux adds additional holes, not called for.

spout. To pour water, he closes the first hole with his thumb and opens the other, under his forefinger; water is then dispensed. He can also pour them simultaneously.

This explains the matter. Here is the drawing.

Figure 27. A. S. 3713, fo. 26b (middle). *Drawing of Chapter 23.* The reference letters are poorly visible, except:
ه (ε): ["Thin Pipe"]
ز (ζ): ["Inner Half of Ewer"]

[Chapter 24. Drink Dispenser with Inlet Siphon and Inlet-Outlet Control by Vents][490]

Construction of another surprising vessel. One prepares a vessel of elegant and very beautiful construction, preferably in form of a ewer, of silver, copper or tin. It has an outflow pipe and a handle. When one pours water into this ewer, while the outflow pipe is closed, no water enters, neither little nor much. When we remove the finger from the outflow pipe, the water enters and the vessel fills. When one closes again, the water is stopped, as before.[491]

---

490 Similar to Hero, I, 36 and Ibn Musa, Chapter 51. Here begins the set of drawings in **A** (Figures 28–54) which has no reflection in the Latin traditions, found up to now.

491 Or rather: it then gradually stops entering, as pressure builts up in the vessel. Also see Drachmann, p. 63.

The construction is the one which we explain here. Take ewer α; close it at the neck by a sheet, centrally pierced by a hole. The neck is long. Insert a tube which goes to the lower part of the ewer, to within 1 finger from its bottom. This tube, γ, projects 2/3 into the ewer's neck δ. Provide another tube, ε, wide enough to contain the first and somewhat more. Let it be long enough to contain the length beyond [the cross-sheet] and somewhat more. Let it be closed at one end, as in medicine capsules, and let its other end be divided in two by a slot, so that it allows entrance of water at the bottom when it is superimposed on said length beyond. This tube [unit] is mounted in the ewer neck as if it were for a vessel discharging when overfilled.[492] The slotted

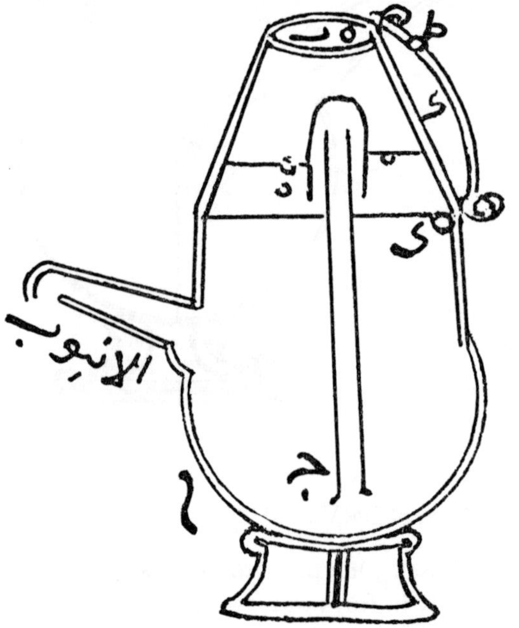

Figure 28. A. S. 3713, fo. 28b (top). *Drawing of Chapter 24.* Reference Letters and Legend:
ا (α): ["Container")
ب (β): "The Tube," Outlet
ج (γ): ["Tube," Inlet]
د (δ): ["Neck"]
ه (ε): ["Wider Tube"]
ط (ϑ): ["Hole closed by the Servant"]

part is soldered on. The ewer head is closed, at a distance of one thumb [above ε], by a sheet, well soldered on to avoid leakage; then it works better. The middle of this sheet has a hole ζ, through which water enters the ewer. The vessel has spout η. This is its construction.

When water is poured into the ewer through hole ζ, it enters and reaches the cross sheet. It then rises in the neck and [outer] tube. When it reaches the line ε,

[492] Chapter 10. Carra de Vaux erroneously refers to Chapter 49.

it returns through the [inner] tube and [reaches] the inside of the ewer. The air then leaves through the spout. This continues until the ewer is full.

If one closes the spout η, the air can no longer pass. Water can then no longer enter,[493] but is stopped and overflows outside. When the orifice is opened, the air leaves and the water enters.

One can also omit an outflow tube on this ewer.[494] The handle can have a hole communicating with the inside of the ewer, and a second [and concealed] hole in the handle surface. The operation then is nicer. The handle is held by a servant, and he who pours water into the ewer will not know the reason for the [change in the] operation. The hole closed by the servant is ϑ; the one into the ewer, ι; the handle $\varkappa$. This is the end of the construction of this vessel. Here is the drawing.

[Chapter 25. Drink Dispenser with Inlet-outlet control by Vents and with Outlet Valves][495]

Construction of another vessel, more surprising than the preceding one. Note that this vessel is a vase called wine-thief.[496] It has this name because of its operation, that is: even when full to the brim, it dispenses only so much as the operator wishes. Although the outlets stay open, it can stop dispensing as if it was empty. Also, if one inverts the full vessel, little flows out. Here is the description.

A pipe is introduced through the mouth of the vase, to a point a finger's width from the bottom. In the middle of the vessel a partition is placed.[497] The pipe is mounted on it. The handles of the vase are curved from the side of the belly to near the top of the convexity of the vessel. They have air holes, [one hole in each handle, going to the inside of the vase, and one] going to the outside, which can be closed with the fingers. Outlets are provided on the two sides of the vessel.

When you wish to fill this vase with water or other liquid [and to surprise the spectators], begin by closing the air holes. The air can then escape only

---

493 Due to internal compression of air. See note 491.

494 See Chapter 25, where outlet faucet valves are used, instead.

495 Hero, I, 22, 23 and Ibn Musa, Fig. 53, suggest the devices of Chapters 23, 25 without their stopcocks. Drachmann, pp. 60–62, seems correct in believing that at least Chapter 25 is post-Heronian.

496 It has little, if anything, to do with the "wine thieves" of Chapters 43 ff.

497 The text does not say whether horizontally (as shown by Carra de Vaux) or vertically (as shown in the next chapter). The present chapter seems to be an unfinished draft.

with liquid, at the [upper] end of the vase. When the pipe is filled to the top [and more liquid is added], the liquid overflows from the neck. To stop such overflowing, the one who holds the vase can slightly open the air holes; depending on such opening, the overflowing stops [and liquid enters]. He can, even when it seems full, reverse it on its mouth; only a small amount of liquid, which was in the pipe, then flows out.

This device also belongs to the control pipe devices.[498]

Let the vase be marked α, the pipe β, the pipe's inlet orifice γ.[499] One closes this air-tight, when the water has been poured [in] and the vase filled. The holes in the two handles are δ, ε; the outlet faucets ζ, η, and the handle holes to the vase ϑ, ι.

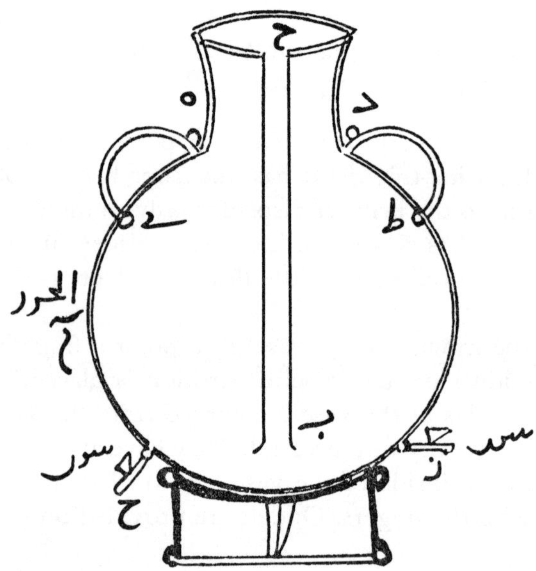

Figure 29. A. S. 3713, fo. 30a (top). *Drawing of Chapter 25.* Reference Letters and Legends:

ا (α): "The Vase"
ب (β): ["Pipe"]
ج (γ): ["Inlet"]
د ه ($\delta\varepsilon$): ["Holes in the Handles"]
ز ح ($\zeta\eta$): "The Faucets"
ط ع ($\vartheta\iota$): ["Handle Holes"]

Water must enter at γ, but if holes δ, ε are closed, it does not enter inside the vessel, as air cannot escape. The two faucets are closed. When one opens the two holes δ, ε, the vase fills. One then carefully closes the orifice γ of the vase, as noted, and opens the two faucets; the liquids then flow [out]. They stop when one closes the two holes again. This can be repeated until the device is empty. Here is the figure.

---

498 As do Chapters 17–24.

499 The latter is mentioned here for the first time. Carra de Vaux adds an overlying plate, with four holes in it. This is not described or authentically shown or technically needed.

[Chapter 26. Second Drink Dispenser with Inlet-Outlet Control by Vents and with Outlet Valves] [500]

Construction of another vessel, similar to the preceding one. If one makes this vessel in another form, it becomes very surprising. For this purpose we can use the form already indicated or other forms; it makes little difference.

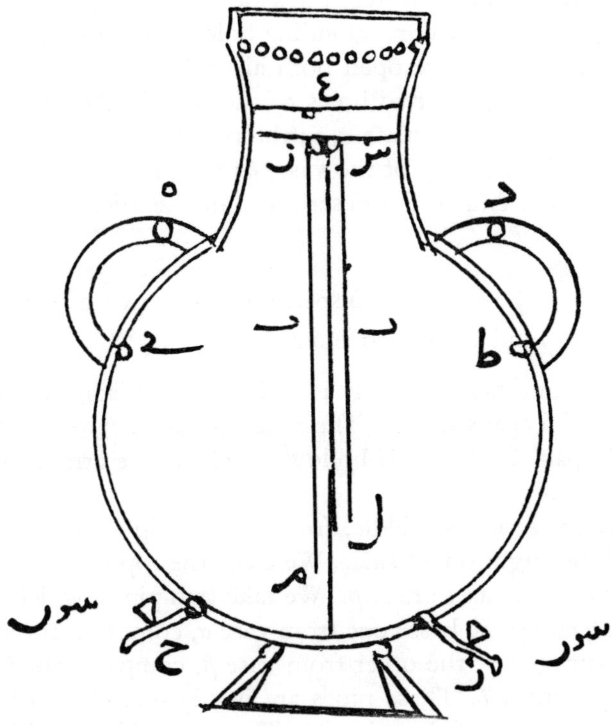

Figure 30. A. S. 3713, fo. 31 a (middle). *Drawing of Chapter 26.* Reference Letters and Legend:

ب ب ($\beta\beta$): ["Pipe"]

د ه ($\delta\varepsilon$): ["Outer Handle Holes"]

ط ع ($\vartheta\iota$): ["Inner Handle Holes (ϑ)"]

ز ح ($\zeta\eta$): "The two Faucets"

ك ($\varkappa$): ["Partition"]

ل م ($\lambda\mu$): ["Holes of the Pipe in the two Sections"]

س ن ($\sigma\nu$): ["Holes below the Plates"]

ع (υ): ["Screen Plate"]

---

500 Substantially identical with Chapter 25, except that the two-part construction, suggested there, is fully shown here. Also see Figure 2e (Deonna, Fig. 6).

We then install the pipe at its proper place; let it be wider than before.[501] We divide the vessel by a partition from end to end, adhering to the wall of the vessel. We put the two halves of the pipe in the two halves of the vase.

We can then use this vessel as the other, or use it with two kinds of liquid. At its mouth, at υ, a "screen" type control plate is needed, with a single hole in it, as described above; we already know this.[502] The plate turns, as do controls in other devices of ours. Water and wine are poured in. To dispense the one or the other, we open the corresponding hole, while closing the other hole. To dispense both together, we open both air holes.

Let the partition be $\varkappa$; the two holes of the pipe in the two vase sections λ, μ; the two holes below the plate σ, ν; the two outer handle holes δ, ε; the inside handle holes in the belly of the vase, ϑ; the pipe $\beta\beta$; the 2 faucets ζ, η; the vase α. This is the end of the description of this nice vessel. Here is the figure.

[Chapter 27. "The Prison:" Dispensing System with two Vessels and Concealed Interconnections] [503]

Construction of another vessel. It is called The Prison. We prepare two vases, on a base of 6 spans length. The vases are at the two ends. The depth of the base is 1 span. Its inside is hollow, to allow the arrangement that we desire.

We divide each vase in its middle, from near its lip to its lower part, by a well-made partition, tight on all sides. We close the top of the front compartment of each vase. The vases are α, β. We take two pipes, which go from one vase to the other, along the base: one from vase α, compartment α, lower part, to vase β, compartment γ; the other from vase β, compartment β, lower part, to vase α, compartment δ. These pipes are in the hollow space of the base, concealed so that no part of them is seen. They are soldered with tin to make the construction reliable and stable.

Let us then take two siphons, with top bend like the hook-knife for pruning trees. Let them extend from the bottom to the neck of the vase. The bend of each siphon is at the throat of its vase, and is bent toward the interior of the

---

501 Probably to facilitate dividing it.

502 Probably refers to hole γ in Figure 27.

503 Somewhat similar constructions, with different details of hollow pedestals and connected equipment, are shown by Hero (I, 14 and 24), Ibn Musa (his Fig. 13), and AlJazari (ed. Wiedemann, Drinking Vessels, p. 78). Drachmann, p. 68 may be right or wrong or partly right and partly wrong when he declares Chapter 27 to be post-Philonian.

vase's curved form. One of them is within [the closed space in front of] partition δ; the siphons are η, ζ. The [lower] ends of the siphons are interconnected by the two pipes concealed in the hollow of the base. Between these two pipes there is a valve, well fitted, which extends outside the middle of the base. It is marked ε.

Let the two vases be mounted, and equipped with two plugvalves, one in compartment γ, the other in compartment δ, where the two siphons are.

Figure 31. A.S. 3713, fo. 33a (top). *Drawing of Chapter 27.* Upper central legend: not deciphered.
Lower legends, right and left: "Faucet"
Lowermost central legend: "The Base." Reference Letters:

ا ب ($\alpha\beta$): ["Vases"]
ه (ε): ["Valve"]
ز ح ($\zeta\eta$): ["Siphons"]

When all this has been finished, complete with mounting of pipes and siphons, soldering the base and adapting it and the plugvalves for the vases, one sees only the vases on the base, and the two plug valves.

On the head of each vase, in the open compartment—for [only] one-half of each vase, the front half, is closed—we provide a plug valve with an outlet pipe. Above it we provide a well-constructed funnel with a sieve. When we wish to pour liquid into the vases, as chosen, we open the plug valve and pour whatever is desired through the sieve-funnel. Thereafter we close the valve firmly to avoid any leak, for this would spoil the operation.

Thus, let us pour water in vase α and wine in vase β; close plug valve ε in the middle of the base; open the vase outlets. No liquid issues, neither little nor much, unless we open the base plug valve. When we open it, the wine comes forth, [but] from vase α, and the water from vase β. When we close again, the flow from the vase plug valves ceases.

One can also pierce another hole in plug valve ε, in the middle of the base, between the other two holes of this plug.[504] When we want that something be dispensed from only one of the two vases, we turn the plug valve accordingly. The vase to be closed then closes and the one to be opened opens.

This is the end of the description of this elegant, surprising, beautiful and wonderful apparatus. The principle of proper operation for it lies in good closure of the head of the the two vases and in soldering their closures to achieve closure of the head of the two vases and in soldering their closures to achieve well what is wanted. The result is very beautiful. Understand it and you will succeed. Here is the figure.

[Chapter 28. Proposed Dispenser for a Fixed Volume of Discharge][505]

Construction of another surprising vessel. We make a vase like those for keeping water fresh.[506] The vessel has a faucet. When we open the faucet, the vase discharges,[507] 1 *ritl* and nothing more, even if the faucet stays open all day and the vase is full. If one then closes the faucet and reopens it, there issues another *ritl*. Each time the faucet is reopened after closure, there issues a *ritl*, not more or less.

The vase is similar to those you know. Inside it one places a chamber holding 1 *ritl*. The chamber is inclined to the side of the vessel.[508] It is made with proper measure. Its inner wall, opposite the faucet, has a pivoted gate, which falls inwards to close; it is well fitted to avoid any leak.

---

[504] Not previously mentioned in this chapter, but probably similar to the valve holes in Chapter 22, or in Figure 1.

[505] Drachmann, p. 68 ("new").

[506] Also see Figure 2, f (Deonna, Figure 15, pp. 229–231).

[507] According to the authentic drawing in **A,** this is an illusion. Nothing discharges, unless door γ leaks, and in that case the entire jar is emptied. The desired operation is possible only if compartment β is vented. Such venting, through the neck of the jar, is shown by Ibn Musa, Chapters 22, 23, and probably was shown in Philo's original, but was lost in the Arabic prototype.

[508] This remark may be a fragment from an original edition, which had the required vent.

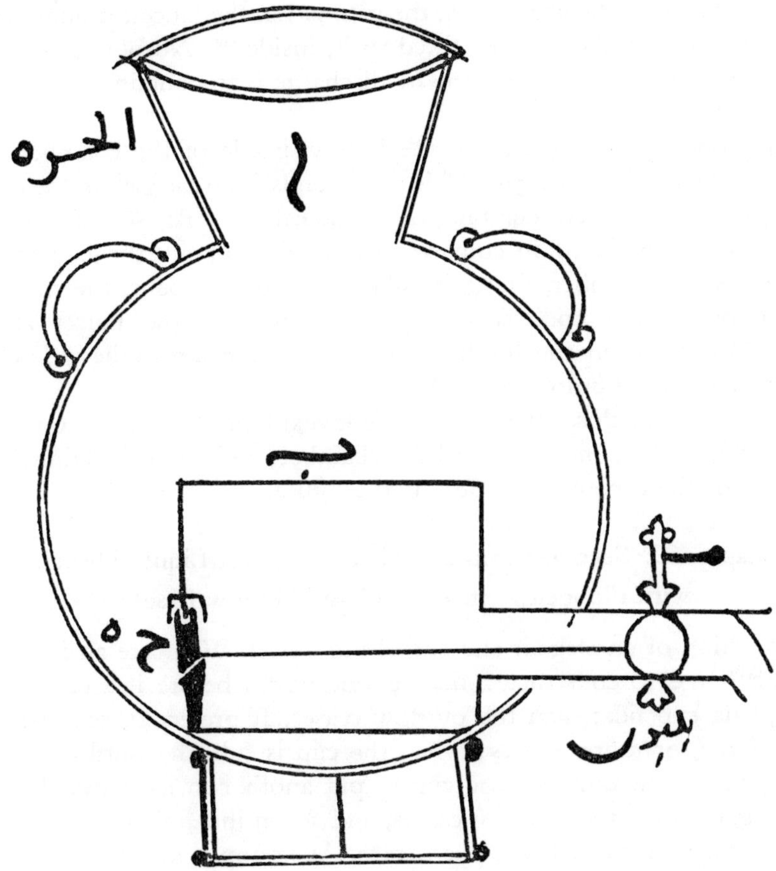

Figure 32. A. S. 3713, fo. 34a (bottom). *Drawing of Chapter 28.* Legends and Reference Letters:

Upper left: "Container"
Lower right: "Faucet"
ا (α): ["Vase"]
ب (β): ["Chamber"]
ج (γ): ["Gate"]

Let the vase be α, the chamber β, the gate γ. Let the faucet, mounted on the vase, go to the chamber and be fitted to it, inside.[509] Nothing from outside this chamber shall reach the faucet. Its discharge is to contain only what is in the chamber.

A thin copper rod is placed in the hollow inside of the chamber, and is attached to the inside of the gate. This rod extends from the gate of the chamber to the plug of the faucet. The plug of the faucet has a flat surface cut into it. When the faucet is closed, its convexity pushes the rod out; the gate opens and water enters the chamber, filling it. When the faucet opens, the flat surface comes opposite the rod, which then enters into the faucet aperture; the gate closes. When the liquid contents of the chamber have left, while nothing enters, the outflow ceases.[510]

If one then again closes the faucet, its convexity pushes the rod out to open the gate. Water then enters to refill the chamber; and so on indefinitely. This is the end of the explanation. Here is the figure.

[Chapter 29. Second Form of a Constant Level Liquid Dispenser with Flooded Vent Control and Hollow Base][511]

Construction of an elegant and surprising vessel. We make an intermittent fountain.[512] It is so constructed that, if one places before it a cup of 1 *oque* volume, this cup fills; then the outflow ceases. If preferred, one uses a cup holding 1 *ritl*, or more or less; when the cup is full, the outflow ceases. It rebegins when one removes the cup to put another in its place.[513] The cup is filled again; then the outflow ceases, and so on indefinitely.

One uses a reservoir shaped like a vase. At its top is a tight cover. In this vase we put an air pipe, which we let enter through the bottom of the vase and which passes through its support and the interior of the base. This pipe is marked α. It rises [at its other end] to the disc β: we place a disc, at β, on the head of the air pipe, turning one way to open and the other way to close. This

---

[509] Perhaps to facilitate its support.

[510] This situation can occur only in the presence of a vent.

[511] Drachmann, pp. 56–60, translating and discussing Chapter 29, rightly questions its operability. Evidently this is a version of Chapter 17, made by an Arabic editor who recognized the problems of that chapter but produced another problematic chapter.

[512] This term is rare in **A,** but also occurs in Chapter 30. It indicates that Chapters 29 *and* 30 are of Arabic origin.

[513] In the four ancient versions, Chapters 17–20, it rebegins when liquid is removed from the cup (by a ladle in Chapters 17–19, or by suction of a wick in Chapter 20).

disc is pierced in its middle by a small, oblique hole.[514] In order to open, the disc is turned to make its hole coincide with air pipe β; to close, one turns to remove the holes from one another and the air hole then closes.

The vase has an outlet opening.

The disc is of measured size. It is somewhat concave to serve as support for the cups placed on it. We take cups of different sizes: small and large, to provide the following arrangement, which you must understand. One cup is

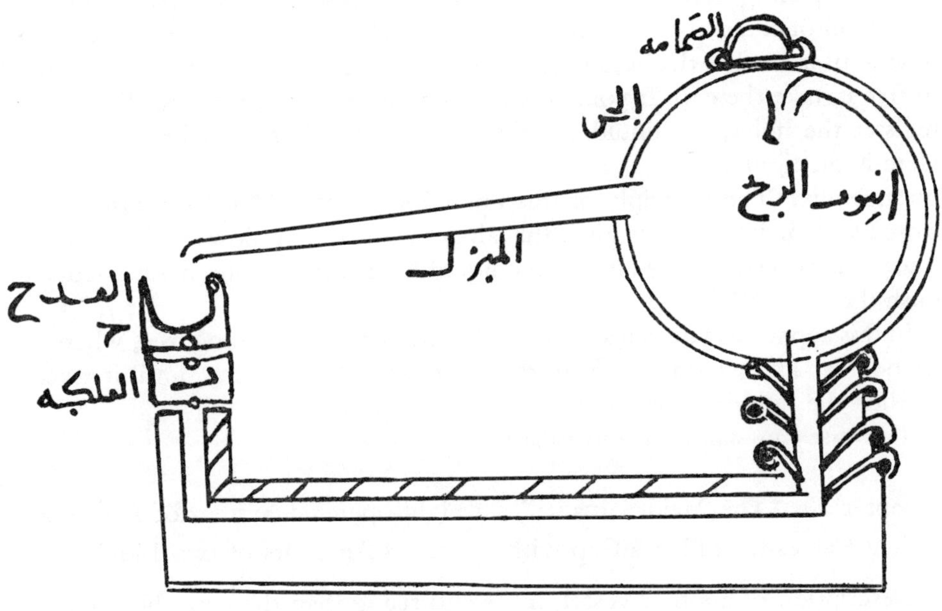

Figure 33. A. S. 3713, fo. 35b (bottom). *Drawing of Chapter 29.* Legends and Reference Letters:

Left top, "The Drinking Glass"
Below the Glass, ح (η): ["Small Hole"]
Below the Hole, ب (β): ["Disk"]
Bottom center, "The Base"
Above the Base: "The Passage"
Right top: "The Plug"
Left of Circle: "The Jar"
Inside Circle, ا (α): "Air Pipe"

in the hollow of another, and there is air between them. The lower cup is pierced at one point. When it is placed on the disc, its hole coincides with that of the disc in open condition, and that of the disc [as mentioned] coincides with the one of the air pipe. The two borders of the cups are soldered together and under the seam, inside, there is a hole as wide as one-half of the little

514 Meaning: the hole is centrally located, in top view of the disc, but slopes to eccentric position on bottom of the disc. So also Drachmann, pp. 58f.

finger. (Or let the lip of the inner cup be below that of the outer one. Bend the outer one over the inner one, without soldering, only extending over it, so that, although air circulates between the two cups,[515] the outer lip seals the inside as if both cups were only one.)

To operate this appartatus, place the cup on the disc; then turn the latter a small distance to place the holes opposite one another. The reservoir has been filled with liquid; the cover has been closed. The liquid pours from the outflow opening because [and so long as] air passes, part for part, from the lip region of the cup to the air pipe opening in the vase. It does not cease to flow until it reaches the hole in the bent lip region; then it is arrested. In order to free it, take the cup. If you, while the cup is removed, turn the disc, to unregister the holes, the liquid remains stopped.[516] This must be understood. Here is the figure.

Appendix. The principle of this operation is this: When the disc hole is closed and air is in the air pipe, this air prevents the outflow of the wine, because it is at rest, for the air and the liquid hold together and neither tends to push the other.[517]

In this apparatus you can also use a clay cup with a rim going to ϑ, where the air hole is. The wine flows from the outlet into the clay cup, until it reaches ϑ; then the hole is closed and the outflow ceases.

The vase can stand on four pillars.

[Chapter 30. "The Maid-Servant:" a weight-operated System of Levers and Vent Valves for filling a Cup with measured Amounts of two Liquids][518]

Description of another vessel, more surprising than the preceding one. An intermittent fountain in form of a maid-servant, who holds a ewer in her hand. If one places a drinking-cup on the palm of her left hand, she pours the desired amount of wine, and thereafter water for mixing with this wine.

---

[515] The Arabic text says, erroneously: "between the two lips".

[516] As noted by Carra de Vaux, the text is "somewhat brief". It is unintelligible.

[517] This seems to imply that, with the disc holes open, the atmospheric air supports the interior liquid: an interesting survival of the view of Empedocles (which also was noted by Roger Bacon and another western Scholastic, at the probable time of the Arabic translation).

[518] As noted by Carra de Vaux, this chapter is copied, in substance, from AlJazari. Details are noted by Wiedemann-Hauser, ed. of AlJazari, Drinking Vessels, pp. 62–93, where a number of variants are also shown. A western Renaissance variant occurs in Palat. Lat. 767, Florence, p. 5; it is reproduced by Prager-Scaglia[77], p. 215.

Make a maid-servant in standing position, of copper or silver. In her head or chest provide a reservoir, and let it be divided into two halves by a partition. For each reservoir [half], provide an air pipe and a liquid pipe to the ewer. The [liquid] pipe of the wine compartment goes directly to the ewer; that of the water compartment is longer and goes around the reservoir, in the belly of the maidservant. The air pipes have inlet openings below the reservoir in the belly of the maid-servant.

The left arm is mounted on two bearing pins, by its shoulder.[519] Inside the figure is a cross-rod, bent at its lower end like a hook-knife for pruning trees, with two rods on it, for two plug valves which have two slots or holes at their ends, to fit into the ends of the air pipes, where they can turn with light friction. The lower end of the cross-rod, similar to a pruning hook-knife, is slightly weighted, to naturally move it down and thereby to raise the left hand.[520] This hand then rises, and the two small plug valves in the pipes then are turned, unregistering their holes from the air pipe holes to close these and thereby to stop the liquid flows through the openings of the ewer. All of this mechanism is inside the maid-servant.

The water and wine are stored in her head, and flow out easily [when the corresponding air valves are open]. The skull serves as a very tight cover. The right hand stays in place without motion; so does the ewer. Two holes connect the two reservoirs with the ewer, as noted.

The description of the maid servant follows. $\alpha\beta$ are the two reservoirs; $\gamma\delta$ the two air pipes; $\varepsilon\zeta$ the two liquid pipes to the ewer; ϱ the bearing pins of the hand ϑ; $\iota\varkappa$ the two plug valves. There is a hole in the figure which opens to the outside.[521] The cover is closed during the operation, in order that air can enter.[522] Otherwise the device would not operate.

When you have completed the above-described figure, take a cup holding a *ritl* or half-*ritl* or any desired volume, fitting the dimensions that you have given the orifices. The mixture should be composed in thirds, for example $^2/_3$ wine and $^1/_3$ water. The size of the cup corresponds to the liquid as proportioned, and the cup is weighted down by a suitable weight. When the cup

---

[519] Actually, several bearings are needed for the required, parallel motion of the hand. The illustration of AlJazari[518], pp. 71f. and 81–87, is considerably better than the reconstruction of Carra de Vaux, but the description of AlJazari is as naive as that of **A.** There is no illustration of Chapter 30 in this manuscript.

[520] These sentences and their equivalent in AlJazari[518] are so vague that a clear, graphic reconstruction of the mechanism is impossible, even in the presence of AlJazari's drawings.

[521] Evidently to admit air to the inlets of the air tubes.

[522] Probably meaning that it can enter only through the air valves.

is placed on the palm of the left hand, the hand sinks; the cross-rod moves; the pipes with plug-valves rise, the [one with the] air-hole for wine before the other, so that its valve [plug] reaches its [open] position more promptly; the [air valve] hole for water reaches its own only when the cup has weighted down [the hand] with [enough initial travel for considerable discharge of] wine. Almost all of the latter then already has been poured. The hand then sinks further; the water[-controlling] plug [air-]valve then rearches its [opened] position, while the wine [control] valve rises beyond its own, closing its pipe and preventing further outflow of wine. The water then begins to flow.

When the figure has poured out what it should, take the cup away from it. The hand then returns to its place, closing the two air holes.[523] Nothing more then comes from the ewer.

If one then replaces the cup in the maid-servant's hand, after emptying it, the hand sinks again; the wine, followed by the water, begins to flow again into the cup. This goes on so long as there is wine and water. So much for this intermittent fountain in form of a maid-servant. Let the device be understood.

Here is the figure. [It is missing in the manuscript.]

[Chapter 31. Bucket-operated System of Levers for Dispensing Measured Amounts of Pumice-Stone and Wash Water, with Apparatus for Initial Hydraulic Delay][524]

Construction of another surprising vessel: a wash stand, constructed after the fashion of [water] clocks [with time-counting balls].

[A structure with] an arch is built, and under the arch, a door opens. A hand appears, holding a ball full of pumice-stone. One takes the pumice-stone and discards the ball. The hand withdraws and the door closes. Water then comes from the beak of a bird [on the structure]. One washes one's hands; there is enough water to rinse both hands and mouth. The outflow then ceases. Then the door opens again; the hand appears with a new ball; one takes and discards; the hand withdraws; water again comes from the beak. These actions are

---

[523] But inherently first passing through their open positions, and causing more outflow of both liquids. Since no cup is receiving the outflow, at the time, this is a messy operation (as is also found in other ill-developed chapters of AlJazari[518] or others, for example Chapters 14 and 31).

[524] Related to AlJazari[518]; however, the drawing in **A** differs from the preceding one of AlJazari, in that it attempts to use perspective. The technology seems to be derived from the bell works of early medieval water clocks, Drachmann, p. 70. About the pumice-stone products used here: E. Schmauderer in *Technikgeschichte* XXXIV, 1967, pp. 300–317 and XXXV, 1968, pp. 205–222.

repeated until everybody has washed his hands; then one closes a valve to stop the water, and the entire mechanism then also stops.

Take a box of desired size and divide it into two halves. In the lower half install a rod on two bearings, with a spoon[-shaped bucket] at one end and a counter-weight at the other to raise the spoon when empty. When slightly filled, the spoon descends and its water pours into a vessel of any desired form in the bottom of the box. Let a water reservoir be placed [above this vessel] in the upper one-half of the box; let it be made of very solid lead, constructed carefully to avoid leakage of water. This reservoir is in back of the box and is provided with a pipe on the middle level of the box, wherethrough the water flows into the middle of the spoon when the latter is empty.[525] The reservoir occupies one-quarter of the box in width and almost the entire box in length. The rod is α; the spoon β; the counterweight γ; the water receiver-transmitter ε; the water supply reservoir ζ; and the pipe η.

Let another rod on 2 bearings be placed in front of this reservoir, in the free space next to the pipe [at right angles to the first rod]. The front end of this rod has a hand extending from it; the other end has a lead counter-weight. The hand rises when nothing is in it. Its weight is such that, when one adds the ball and its contents, the hand sinks. It then presses against the door, which has 2 leaves; the hand then appears on the outside. When one has taken the ball from the hand, the latter returns backwards and re-enters. It then pulls two strings, attached to the two door-leaves, which then close again. The rod is ϑ; the hand ι; the counter-weight $\varkappa$; the two strings λ; the two door leaves μ.

Above the hand one disposes a somewhat inclined channel for the balls, under the arch, leading down to the hand. The balls are [thereby to be] placed in it. At the lower end of the channel there is a gate, which opens naturally when weighted down [by the balls]. A string is attached to this gate, passes over a small pulley, and has its other end fixed to the counter-weight at the far end of the spoon rod. The gate is σ; the pulley ϱ. The string will be described now.[526]

When the counter-weight sinks, it draws the string over the pulley and closes the gate. When the spoon is full and sinks, it relaxes the string and opens this gate; a ball then rolls into the hand, which then sinks, opening the door-leaves so that the hand appears.

Near the water reservoir there is an inlet structure. It is outside the box and comprises a funnel with a valve. When the valve is opened, water flows

---

525 As in Chapter 30, the water is simply spilled at other times.

526 Carra de Vaux: "As to the string, this is noted above"[?].

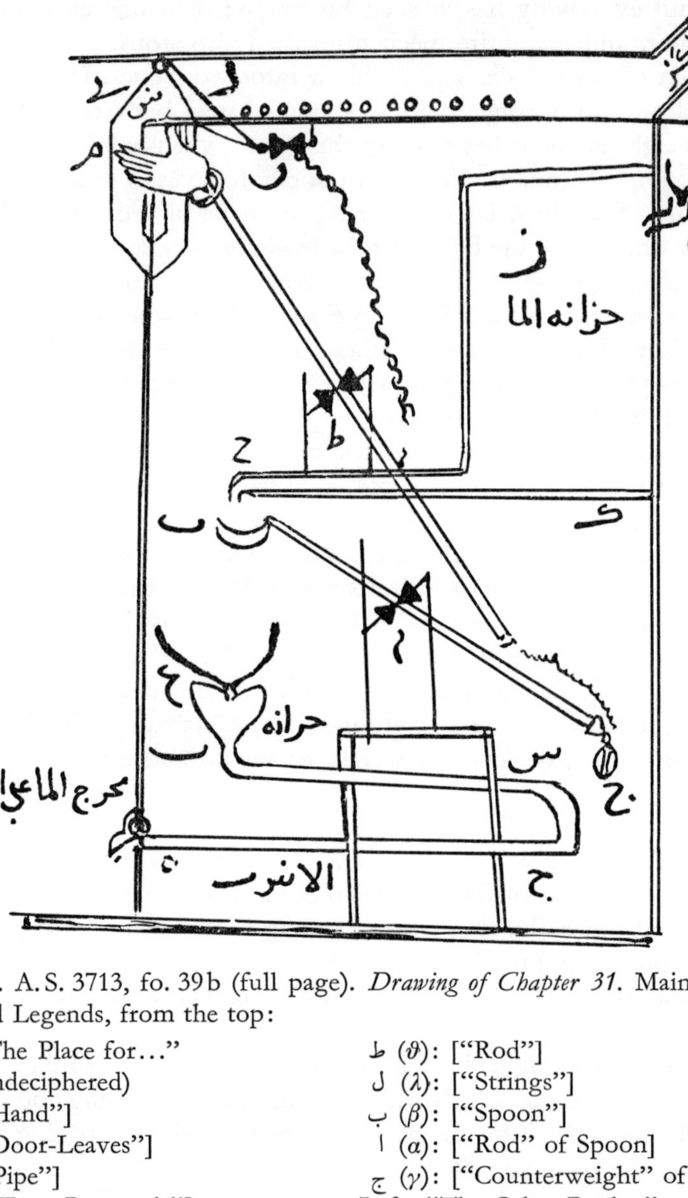

Figure 34. A. S. 3713, fo. 39b (full page). *Drawing of Chapter 31.* Main References letters and Legends, from the top:

Right: "The Place for..." (undeciphered)
ع (ι): ["Hand"]
م (μ): ["Door-Leaves"]
ح (η): ["Pipe"]
ز (ζ): ["Water Reservoir"]
ط (ϑ): ["Rod"]
ل (λ): ["Strings"]
ب (β): ["Spoon"]
ا (α): ["Rod" of Spoon]
ج (γ): ["Counterweight" of Spoon]
Left: "The Other Bucket"

into the reservoir, which it fills to the level of the funnel. The valve is left open. The water drips into the spoon, so long as the valve is open.[527] In the apparatus built as described, the vessel into which the spoon pours also resembles a spoon or a bucket.[528] It is inside the box. In its bottom is a hole, through which the water passes into a long, curved transfer duct to the outside of the box. This curved duct is arranged to discharge the water only after a delay, during which water is retained in the other bucket. Under the hole of that bucket is the other receiver. The water first enters into this second receiver with the speed [corresponding to the size] of the first bucket's hole, and then issues from this second receiver.[529] It issues by a straight pipe going to the curved pipe. The [middle] bucket is v; the lower one φ; and the [curved] pipe σ'. [Thus] the upper bucket provides a vessel which receives some water, then discharges it through its outlet to allow moistening of the pumice stones for rubbing the hands, and only then passes the mass of the water. It all flows into a place o.[530]

Let this description be understood. When the apparatus shall not operate any more, close the funnel valve; no water will then continue to drip. This is the device that we wanted to describe. Here is the figure.

[Chapter 32. Simple, Weight-operated Lever System for Dispensing Wash Water]

Description of another vessel, again for a wash stand. It is of easier construction than the preceding one, wherefrom it is, incidentally, derived.[531]

Take a box of suitable size, with a vaulted top, as shown. Divide it vertically into thirds. In the upper part, below the cover of the box, provide a water reservoir, carefully sealed with leak-avoiding tar, to let water flow therefrom [only to the desired point]. The box is α; the water reservoir β.

In the box, erect 2 columns, a cross-beam, and on it a scale-beam. The columns are α'; the cross-beam β'; the scale-beam γ. One end of the scale-beam has a hand fastened to it, with a hole in it. The hand is δ. Take a cup of 1 *ritl* or more, with a pin on its bottom entering [the hole in] the hand to hold

---

527 Evidently the water-ducts are narrow, thereby providing this "dripping."

528 This vessel is not said to tilt. Perhaps it is just a funnel which first discharges through its spout and then overflows.

529 This evidently happens after overflowing the middle "bucket," indicated in the authentic drawing.

530 The contents of this last paragraph seem to be an afterthought or an interpolation.

531 Derived recently? by simplification?

it well. The cup is ζ. On the other end of this beam place a lead counter-weight, which raises the hand and cup when the latter is empty. When the hand is full, it overpowers the counter-weight; the hand then sinks and issues through a door with two leaves, which it pushes open. Two strings are attached to the wrist of the hand, as described for the foregoing figure. The weight is ε; the door ϑ; the two strings η.

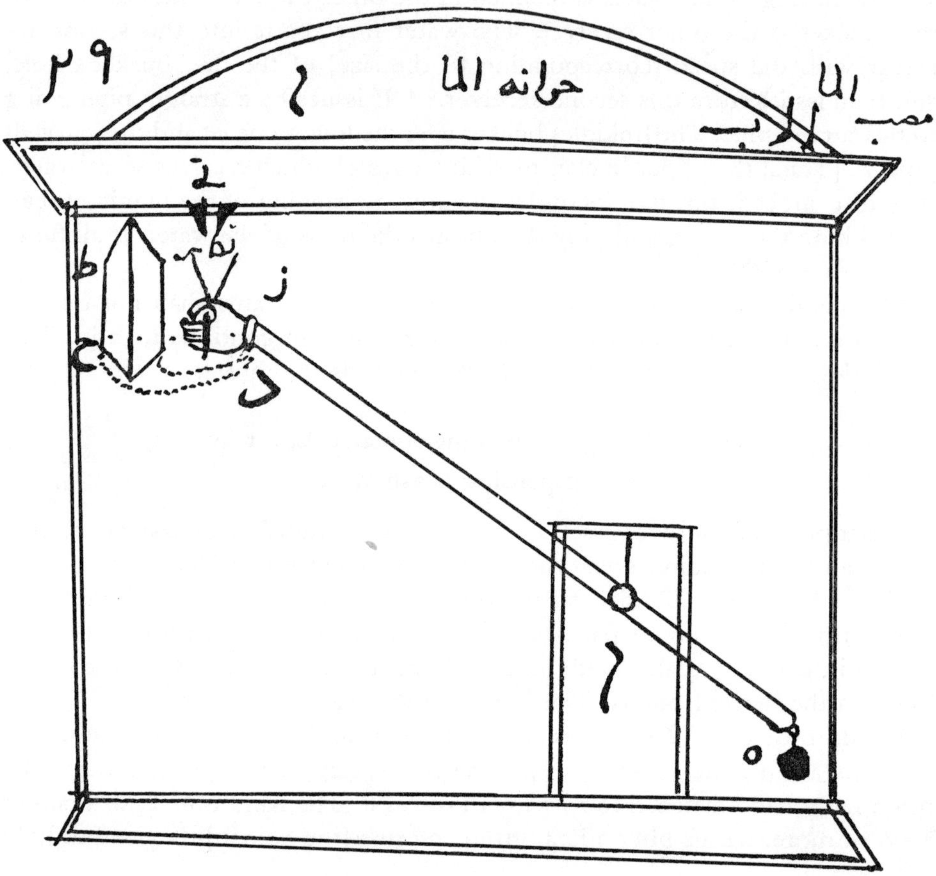

Figure 35. A. S. 3713, fo. 41 a (top). *Drawing of Chapter 32.* References: see Figure 34.

The reservoir has an orifice above the cup. When the cup, in the hand, is empty, the water drips from the hole, drop after drop, into the cup to fill it.[532] The hole is ι.

[532] Again, it also drips when the cup is full. Either a copyist omitted a mechanism whereby the "hand" closed a valve, or the messy operation was simply allowed.

When the cup is full, it sinks down and pushes the door, which opens; the hand appears. Take the cup; empty it; replace it in the hand;[533] then leave it; the hand then returns to its original position. It reappears when the cup is full again. It re-enters when you have taken it and emptied it. This continues until there is no water any more in the reservoir. Let the operation be understood. Here is the figure.

[Chapter 33. Container System with Siphon and Flooded Vent for a Wash Water Dispenser and Sink Control] [534]

Description of another wash stand device. We take a basin, giving it such diameter as we want, for example 4 spans, more or less. Below it is a receiver for waste water. In the middle of the basin we provide a concave portion, forming a hollow chamber. On it, to-wit around it, we dispose 4 columns, whereof 3 are solid and 1 is hollow. In the lower part of this one is a hole, open toward the rim of the container [or basin]. On these columns we place a vessel for water, in form of a pot or vaulted structure or jug. This vessel is soldered to the columns. Within it, an air tube is mounted, and this is connected to the hollow column so that they together form a single conduit. The vessel has an outlet pipe, wherefrom the water flows over the hands. We provide the vessel with a well-fitted cover, tight all around. The basin is $\alpha\beta$; the form of the chamber is as shown.[535] The hollow column and air tube is $\gamma\delta$; the water vessel ε; the outlet pipe ζ.

Let a horse be made of copper, holding his head near the floor of the container, as if he drank the water.[536] Let an oblique passage go from his mouth to his neck and belly, and let it be elbowed, like a hook-kife for tree-pruning, to withdraw the water. A pipe is provided on the belly to communicate with this passage. The end of this pipe enters into an outlet hole in the center of the chamber.[537] Let the pipe of the horse have a hole opposite this outlet hole of the chamber, so that the water may go from there into the waste

---

[533] This had to be held, manually, in the meantime.

[534] The text and drawing of this chapter, and those of Chapters 34 and 35, become technically understandable only by explanations, which are therefore added, in brackets.

[535] The writer seems to admit his inability to describe it. The draftsman, to whom he refers, seems to combine plan view and sectional elevation, in a way which also leaves the subject matter unspecified.

[536] Is the *concavum* open on top, or is it closed or a float (as the Carra de Vaux drawing seems to suggest)? The operation is poorly intelligible in either case.

[537] Probably meaning: in a central outlet chamber, below the hollow support.

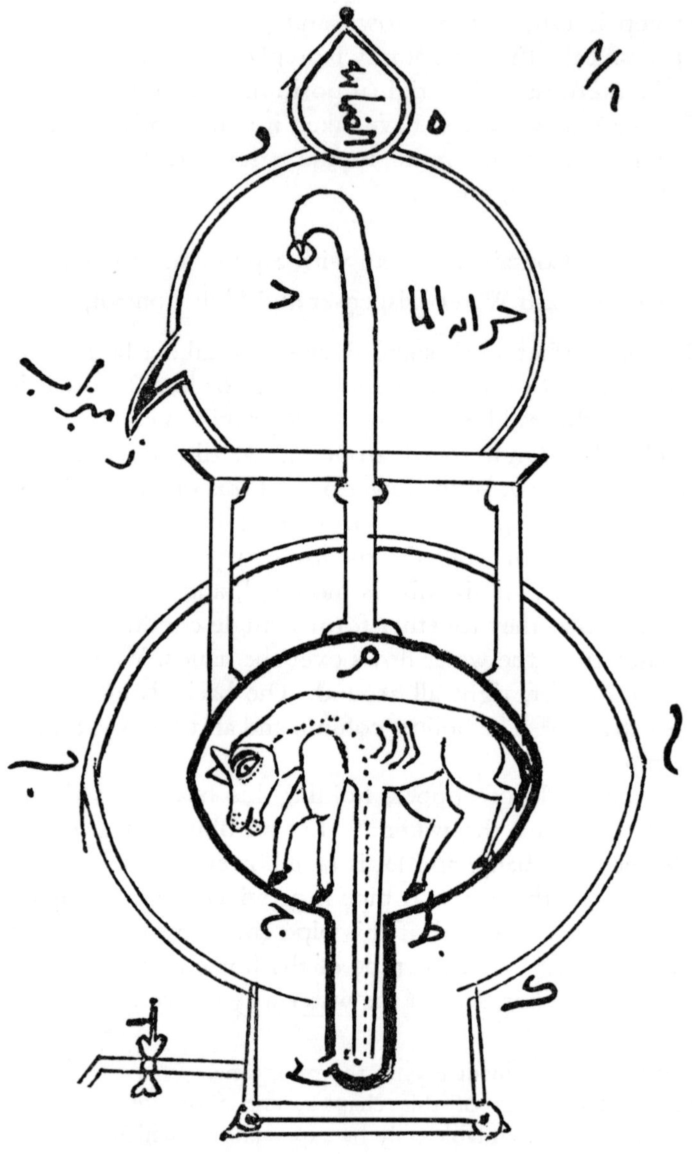

Figure 36. A.S. 3713, fo. 42b (full page). *Drawing of Chapter 33.* References and Legends:

Top: "The Plug"

د (δ): ["Air Tube ($\gamma\delta$)"]

Right of δ: "Water Reservoir"

ز (ζ): "The Outlet Pipe"

ا ب ($\alpha\beta$): ["Basin"]

م (μ): ["Hole"]

ح (η): ["Horse's Mouth"]

ط (ϑ): ["Outlet of Horse's Pipe"]

water receiver below the basin, to be collected in it. The horse's mouth is at η; the outlet end of the horse's pipe is at ϑ, and the outlet hole [of the chamber] at ι. The collecting receiver has a valve $\varkappa$ for tapping off water.[538]

One pours water into the vessel. One closes its cover and [for the moment] its outlet pipe. When one desires one opens [this pipe, while venting already is provided at] the lower end μ of the air pipe:[539] the user connects vent hole μ to the [vessel with this outlet] pipe, to wash his hands.[540] Water then flows on his hands, until hole μ is submerged [and the vessel's vent thus is flooded]. The horse then is ready to drink, as in this condition the water level is as high as the neck of the siphon in the horse. The horse drinks rapidly, as the horse's siphon is wide; the pipe [from the vessel] is smaller by at least one-quarter. This pipe is vented again and resumes discharging. Its flow ceases [only] when the water closes hole μ, at which time the horse [again rapidly] drinks. This repeats itself until the end.[541] Such is this device. Here is the figure.

[Chapter 34. Second Flooded Vent and Siphon Structure for Washing Hands][542]

Construction for another elegant wash stand of surprising aspect, similar to the preceding one. One takes a basin similar to the one just described, with 2 or 4 columns around its chamber, one of them providing an air pipe. The basin is $\alpha\beta$, the air pipe γ. The last hollow column[543] goes from δ, the inside of the [upper] water container, to the inside of the basin at a point ε near its rim. On the rim let a human figure be placed, holding a jug above the user's hands. The figure has a duct through one of its legs and its body and to its hand and the jug. Mount the figure on hole ε so that passage δ in the column communicates with the figure and its jug at ε, forming a single tube for flow of water from the reservoir to the jug. The water outlet of the jug is η, and the figure's hollow leg is ζ. Provide the hollow chamber inside the basin with

538 Shown with unclear reference letters in the authentic drawing, and at the wrong place by Carra de Vaux.

539 In this sentence and the next, the text is repetitive as well as confused. The translation of Carra de Vaux is technically unintelligible. The present translation is very free.

540 Carra de Vaux gratuitously provides this pipe with a plug valve.

541 That is, until all water has left the upper vessel.

542 The notes to Chapter 33 also apply here. The figure is drawn in the Celtic Style.

543 Or better: "a second hollow column." This column and its ornamentation is the only thing added to Chapter 33.

Figure 37. A. S. 3713, fo. 44a (full page). *Drawing of Chapter 34.*
References and Legends: similar to Figure 36.

any desired drinking figure, having in it a tube to the outlet hole of the basin, as the horse. For example replace the horse by a bull. Water enters the bull at ϑ. The end of the [siphon] pipe is at ι. The outlet hole is at $\varkappa$. The large reservoir has a water outlet reservoir, as before, which receives the water drunk by the bull; it is λ. It has a faucet to withdraw water from it. The upper reservoir has a well-fitted, light cover. Let all this be made, as before, and let it be understood.

When the [upper] reservoir is full and the cover closed, shut off the jug and [for this purpose, submerge] opening μ. You can then re-open the jug and [for this purpose, expose] opening μ. The water then promptly flows from the jug, and continues to do so until it [again] reaches up to μ; then it ceases; the bull drinks; it does this rapidly, like the horse. Water flows again; and so on, until the reservoir is empty. Such is this design. Here is the figure.

[Chapter 35. Jet Unit with Gear-Driven Adjustment][544]

Construction of another vessel, a jet device for rose water or other liquids. A housing is made which has a jet device for projecting the liquid through the beak of a bird which "flies" in the desired direction. The construction is simple.

Take a housing, of the hexagonal form shown here. Below it is a reservoir for the liquid; in it, the mechanism. Let the housing be well proportioned and ornamented and stand on nice feet. Let each corner of the hexagon have a copper pillar, surmounted by a flower ornament or other desired figure. The hexagon is shown as $\alpha\beta\gamma\delta\varepsilon\zeta$. Let the pump be placed in one of the pillars, for example δ. Let the middle of the housing have a vertical column, about 1½ span long, surmounted by a bird with unfolded wings. The liquid is squirted from its beak. It turns at will, to direct it toward the desired person and to squirt liquid in his face.

Let one pillar of the housing—pillar δ—have pump device in it, similar to [other] squirt pumps, as at ζ. Let this device have a liquid-impelling piston at η, with a handle in the top of its flower ornament ϑ. At the bottom of this device let there be a reservoir, into which the pump plunges at ι. Let the pump device have inlet passage $\varkappa$ and discharge passage λ. At the end aperture μ of the discharge passage let a pipe be soldered on, the end of which

---

[544] Based on Hero II, 32 according to Hammer-Jensen p. 416 and Drachmann, pp. 64, 100; however, Hero's variant could be based on the original form of this chapter. The Arabic draftsman again, as in Chapter 30, makes an attempt at perspective. He also shows another device, in the upper left, which is not described.

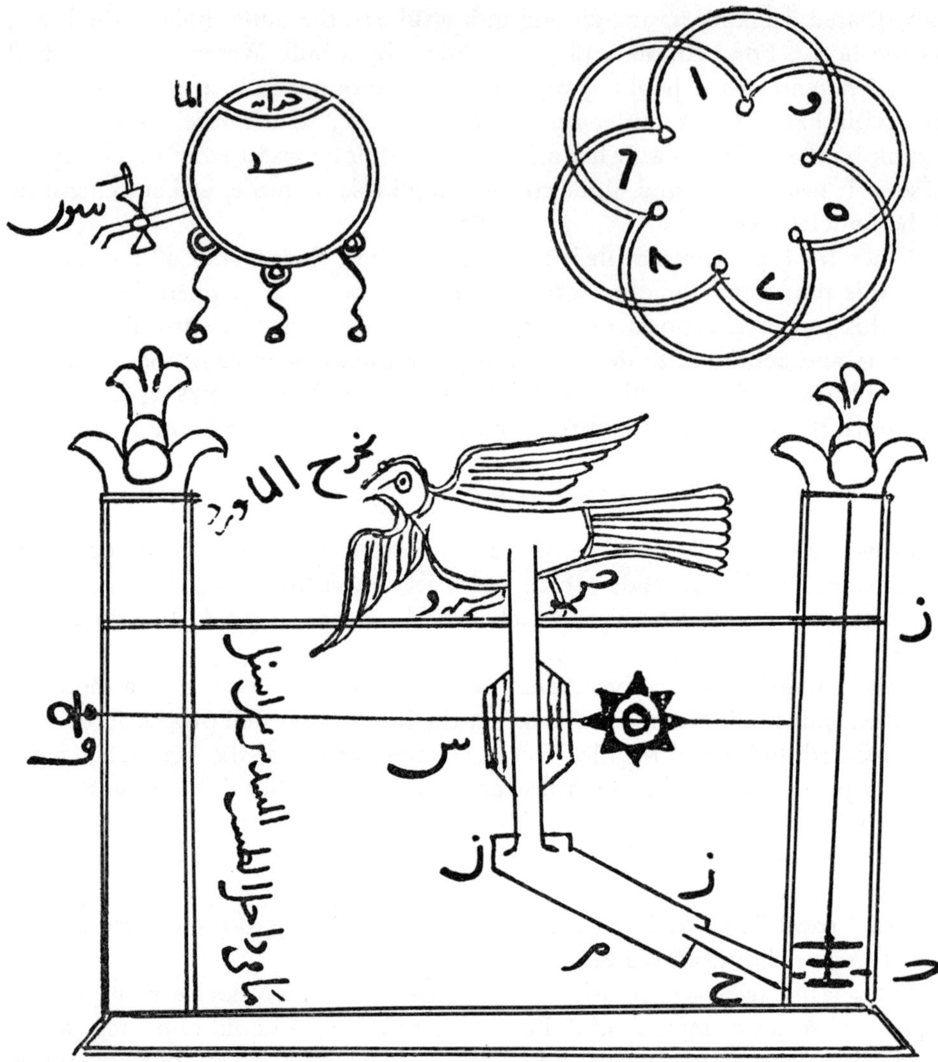

Figure 38. A. S. 3713, fo. 46a (full page). *Drawings of Chapter 35, and unexplained Drawing.*
References in upper right figure: ا, ب, ج, د, ز, ك [?] (α, β, γ, δ, ζ, $\varkappa$). Legends of lower figure (near the bird), above: "This is where it comes out." Below: "[This is] what is inside the Container." Legend and References of upper left figure, at left: "The faucet;" in the circle: ع (ι or υ).

turns up and has a cover ν with a central hole, the elbow portion of this pipe being exactly opposite the center of the housing. A thin, straight tube, with its end entering the hole of the cover, extends from this hole in the center of the housing, and has such proportions as indicated. On a sleeve around it, there is a wheel or lantern-wheel; if it is a wheel, let it be toothed. It is shown at σ. A shaft is provided, carrying a toothed wheel to turn this wheel. Its end, with a handle, extends out of the housing. The shaft is υ, the handle φ.

We also take a bird, σ', with unfolded wings, which has in its beak a tube to its [hollow] interior. We mount the bird at $\varkappa'$ on the tube which has the wheel. If you now want to squirt liquid in the face of someone of the company, turn the handle to direct the beak of the bird toward him. Have the flower ornament with the pump handle before you. Raise the handle and lower it rapidly. The liquid will spring forth and the beak of the bird will project it into his face.

This device is [also] used to perfume garments. Let this be understood. Here is the figure

[Chapter 36. Float Vessel with "Maid-Servant" for a Wash Stand][545]

Description of another elegant vessel. It serves for ablutions. In its middle a maid-servant appears, seeming ready to serve. When one commands her to stop pouring water, she returns to her place.

Take a pan with a deep reservoir below it, to hold enough water for one person or two to wash their hands. Inside this vessel place the figure of a standing maid-servant on a bladder. The pan has a central lid secured to it by a hinge above the head of the maid-servant, to open upwards. Around this lid, the pan forms a screen. The pan is $\alpha\beta$, the reservoir γ, the servant δ, the bladder ε, the lid ζ.

[Another container, around] the reservoir, has faucet η, to dispense the water. When one pours water in the pan, it collects on the sieve [and gradually passes through it]. The float then rises with the servant [who then opens the lid]. Thereupon the water enters the [outer] reservoir through two holes ϑ, which are narrower than those of the sieve [and are no longer sealed by the bladder]. Gradually the servant returns [downward].[546] Here is the figure.

---

545 Drachmann, p. 72, considers Chapter 36 as part of the group of automata which includes Chapters 40–42. Stylistically, Chapter 36 belongs to the Celtic-influenced group, which includes Chapters 33 and 34.

546 Probably the "bladder" also serves as a valve closure for the small holes. This would explain how one can "command her to stop pouring water."

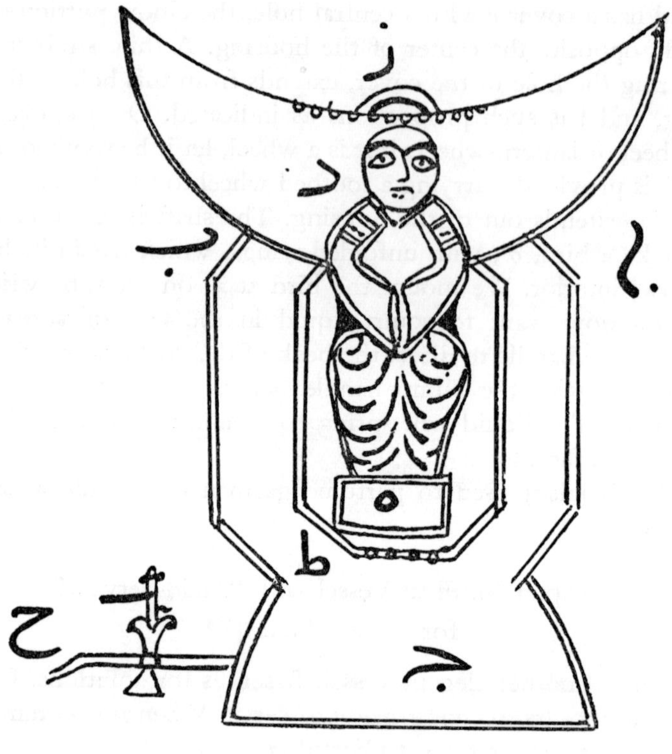

Figure 39. A. S. 3713, fo. 47a (top). *Drawing of Chapter 36.*
References:

ا ب ($\alpha\beta$): ["Pan"]
ج (γ): ["Reservoir"]
د (δ): ["Servant"]
ه (ε): ["Bladder"]
ز (ζ): ["Lid"]
ط (ϑ): ["Holes"]
ح (η): ["Faucet"]

[Chapter 37. Jet Unit]

Description of another ewer of an earlier type.[547] This is a ewer with a jet unit in it. One washes oneself with it. When one wishes, one may squirt water on anyone, as desired.

[547] Perhaps meaning: a unit of the type shown in Chapter 35. Drachmann, p. 63, classes this device and that of 38 as "Arabian" and that of 39 as "Philonian with Arabic improvements." He is silent about the reference to Chapter 13, in Chapter 38, and about the question of successive developments of Ctesibian pumps.

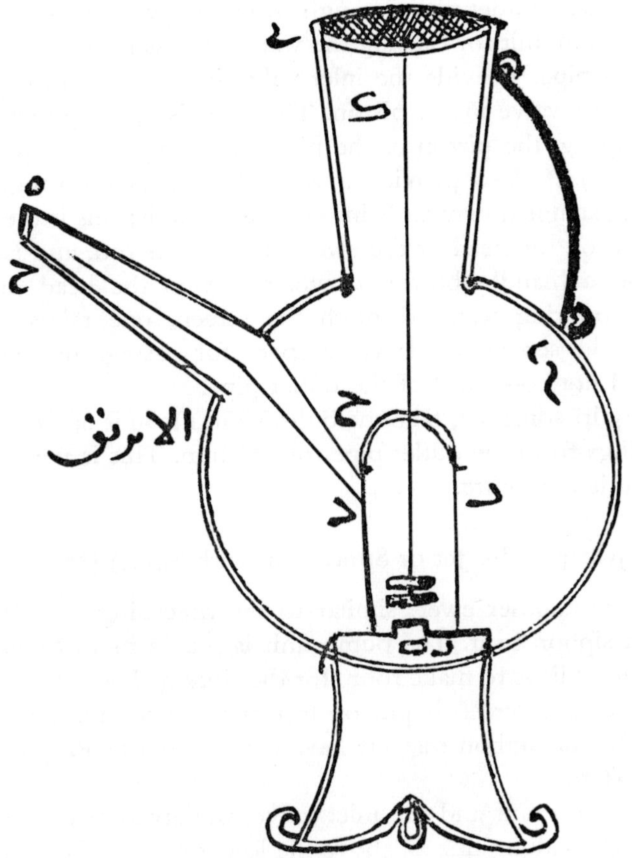

Figure 40. A. S. 3713, fo. 48a (top). *Drawing of Chapter 37.*
Legend and Reference Letters:

ا (α): "The Ewer"
ب (β): ["Pump Unit"]
ج (γ): ["Outlet Pipe"]
د (δ): ["Piston"]
ه (ε): ["Piston's Jet Pipe"]
ز (ζ): ["Inlet Valve" of Pump Unit]
ح (η): ["Outlet Valve"]
ط [?] (ϑ): ["Piston Rod"]
ع (ι): ["Button" Handle of Piston Rod]

Take a ewer, as the one commonly known, wherein you place the jet pump unit. The ewer has an outlet pipe through which the water is dispensed. Place a piston in the pump unit for causing the water to pass into a thin pipe which enters the ewer's pipe. Provide the inlet valve in [the bottom of] the pump unit, and the outlet valve in the piston. The ewer is α, the pump unit β, the ewer's outlet pipe γ, the piston δ, the piston's jet pipe ε, the inlet valve ζ, the outlet valve η.[548] Also provide a rod which passes through the neck of the ewer and extends downwards into the pump unit; this is the piston rod ϑ. Place a screen on the head of the ewer, around this rod, and fasten to the top of this rod a [handle-]button ι, which covers the head of the ewer entirely.[549] For pouring water through the screen, one raises the button. Between it and the screen one leaves a space for passage of air; otherwise the water would not come out of the outlet pipe.

In order to squirt someone, take the button and move it up and down. The water then springs from the outlet pipe toward him. This is what we wanted to explain. Here is the figure.

[Chapter 38. Jet or Spray Unit with Siphon][550]

Construction of another ewer, similar to the preceding one. We use this ewer and add a siphon to it. The pump unit is put in its place as described, but is moved back a little to make room for the siphon. The ewer is designated by the same letters. A screen is put on top of the ewer and [another] in its lower part, as for the siphon that we have constructed in Proposition IX of our present book[551].

Between the pump unit and the inlet valve, the apparatus is divided by a sheet, which extends along line $\varkappa\lambda$ above the lower screen of the ewer. In this sheet one makes a hole ε, wherein the bent siphon is mounted. The siphon's upper part or elbow portion σ is in front of, and close to, the entrance of the outlet pipe. The water level never is above the elbow of the siphon, in the operation of the vessel.[552] A mark is made to show this level.

One can wash and squirt with this vessel in the same way as described. To remove all water from the ewer, we blow strongly into the outlet pipe at γ;

[548] The pump valves are vaguely shown in the authentic drawing.

[549] Apparently to allow the operator to demonstrate that, with the "button" closed, water cannot be poured from the spout. See the next two sentences.

[550] See note 547, about the relation to Ctesibius.

[551] That is, in Chapter 13 (see Chart 5).

[552] Literally: "when the vessel is stirred." As in Chapter IX (now 13), there always is air at this "entrance" and thereby the water-squirt can act as a spray.

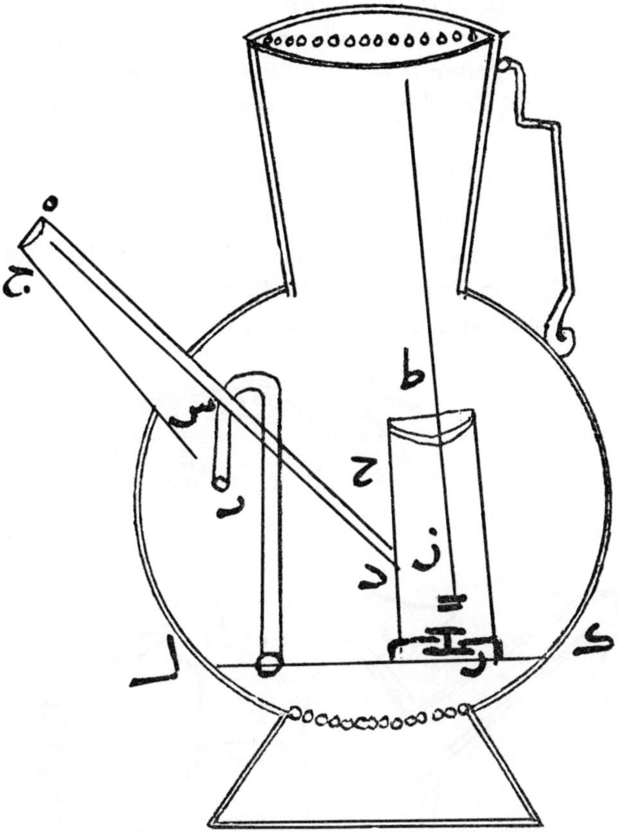

Figure 41. A. S. 3713, fo. 49a (bottom). *Drawing of Chapter 38*. Reference Letters:

ج (γ): ["Outlet"]
د (δ): [Pump Unit, not described]
ط (ϑ): [Top of Pump Unit, not described]
ح (η): [Pump Outlet, not described]
ز (ζ): [Piston Rod, not described]
upper ه (ε): [Jet Pipe, not described]
lower ه (ε): ["Hole" for the Siphon]
ك ل ($\varkappa\lambda$): ["Sheet"]
ر (ϱ): [Siphon Inlet, not described]
س (σ): ["Elbow" of Siphon]

the water then enters orifice σ of the bent siphon. It reaches the elbow and then all of the water[553] leaves, as one part draws the other. When we draw out the air, the water is drawn from the elbow of the siphon and reestablishes the former condition.[554] Let this be understood. Here is the figure.

[Chapter 39. Vessel to be Filled only by Immersion][555]

Construction of another surprise-ewer. This vessel is light and elegant. One makes it of clay, copper or lead or other desired materials. When water is poured in it from the top, it leaves at the bottom and refuses to stay inside. It fills only when dipped and immersed in deep water.

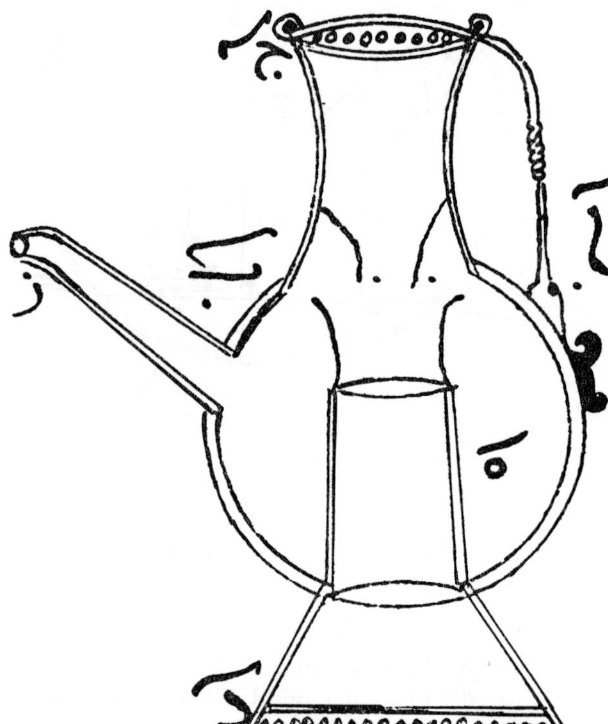

Figure 42. A. S. 3713, fo. 50a (top). *Drawing of Chapter 39.* Reference Numerals:

آ ب ($\bar{\alpha}\bar{\beta}$): ["Vessel"]
ج ($\bar{\gamma}$): ["Top"]
د ($\bar{\delta}$): ["Bottom"]
ه ($\bar{\varepsilon}$): ["Tube"]
ز ($\bar{\zeta}$): ["Outlet Pipe"]

Take a vessel as shown, with a screen on top. In the middle is a vertical tube. The vessel has an outlet pipe, as ewers have. The vessel is $\alpha\beta$, the top γ, the bottom δ, the tube ε, the outlet pipe ζ.

When water is poured into the vessel, it leaves at the bottom, as the screens overlie one another, but when it is dipped in water, this enters and remains around the tube. Let it be understood. Here is the figure.

[Chapter 40. Birds and Snake as Clepsydra Clock Parerga or Automaton][556]

Construction of another surprise vessel. Let us now describe how one makes a covered reservoir, surmounted in its middle by a tree with branches, and a bird thereon, hatching its little ones. A hole is formed at the root of the tree. When one pours water or wine in this receiver, the liquid enters the inside; then, from the hole at the root of the tree, there leaves a snake, which rises and frightens the little birds. When the snake's head is near the chicks, the bird jumps, rises, and flutters with its wings. When the water then leaves the reservoir, the snake goes back down. When it has re-entered its hole, the bird comes down on its little ones again, folds its wings, and resumes hatching.

One takes a reservoir of a diameter of 1 span, more or less, as is desired by the builder and depending on what it shall contain. You give this chamber a cover, such as a copper plate, arranged as a screen, which can be formed and decorated as you wish. In the middle of this screen erect a tree, with its foot soldered to the screen surface. Its trunk is a hollow tube; it carries branches, leaves and fruits as desired. The receiver is α; the screen β; the tree γ.

In the hollow space of this pipe, another pipe is introduced, which reaches the inside of the reservoir and has a float on its lower part, floating on the

---

553 Down to the siphon intake. (The statement is copied from Chapter 13.)

554 Meaning: we re-establish a water level at the siphon intake. Of course we must also stop pumping.

555 For ancient vases from Cyprus and from Etruria, similar to the present one: Deonna, Fig. 1–3, pp. 210–213. For Arabic variants: Ibn Musa, drawings 4, 14, 15. On the drawing in **A,** most of the references to this chapter are inserted as numerals. Perhaps the Persian prototype had omitted them, and an Arabic draftsman added them after the original translation. Also see note 556.

556 The drawings of Chapter 40–42 rely, more largely than is usual in **A,** on legends instead of reference letters. The entire set of automata thus appears to be from another book (of Philonian origin) than the original *Pneumatica.* The descriptions of the automata are more elegant and more finished than the preceding and following chapters; they seem to have suffered less abbreviation.

water at the neck of the snake; it is constructed like a clepsydra float. The [inner] pipe is hollow from one end to the other and is well soldered and adjusted so that the water may not pass it. The float has a hole in form of a tube, of suitable width.[557] The pipe is δ; the float ε; the hole ζ.

Next, we make a thin silver snake, of coiled silver wire or light hollow sheet construction. It fits in hole ζ of the float. The snake is centered on a small float [of its own], which also rests on the water, while supporting the animal. The snake is η; its float ϑ. The large float which carries the pipe inside the [stationary] tree pipe is at one-third the depth of the reservoir, from the top. This pipe and float do not descend downwards more than that. The head of the pipe is at the level of the head of the tree.

We then take a bird of the desired kind. It is hollow and is soldered to this pipe; the hollow of the pipe joins the hollow of the bird. The bird is ι.

The snake float rests on the bottom of the reservoir; the water begins to raise it as soon as the reservoir begins to fill. The snake then comes out and rises toward the chicks. The water does not yet reach the float of the bird; this requires an interval, calculated from the length of the snake, so that the water reaches the bird float when the snake is near the chicks. As the water continues to rise, the bird float [then also] rises. It [soon] reaches the limit of its path, causing the bird to open its wings.

Let us describe the wings. The back of the bird is slitted by incision, as shown in the figure at $\varkappa$. You make a thin, solid rod, which extends through the bird pipe; its lower end is fixed to the bottom of the reservoir. The length of this rod is such that its top end is at the exact hight of the bird's back, in the slot at $\varkappa$, when the bird sits on its place on the chicks. This end of the rod has a ring soldered to it; the ring has a pivot bushing, also soldered to it. The wing pivots are connected between the two wings, through the ring of the rod. Each wing pivot has a bolt going through this ring, with both bolt ends flattened, so that the wing pivots move easily about the bolt, the latter being thinner than the wing pivots and letting the wings move easily.[558]

When all this is prepared and the bird rests on its chicks, the wings are loose and closed. When the water then reaches the bird float and raises the bird, he raises and spreads his wings, as they [in the slot] are fixed to the rod, as described. The rod is λ; the wings μ.

When all this is executed as described and when we have assembled it, water is poured on the screen and enters the reservoir bottom. The snake float rises

---

[557] Evidently the inner pipe is soldered to the top and bottom sheets of the float.

[558] The bolts evidently are fixed inside the bird.

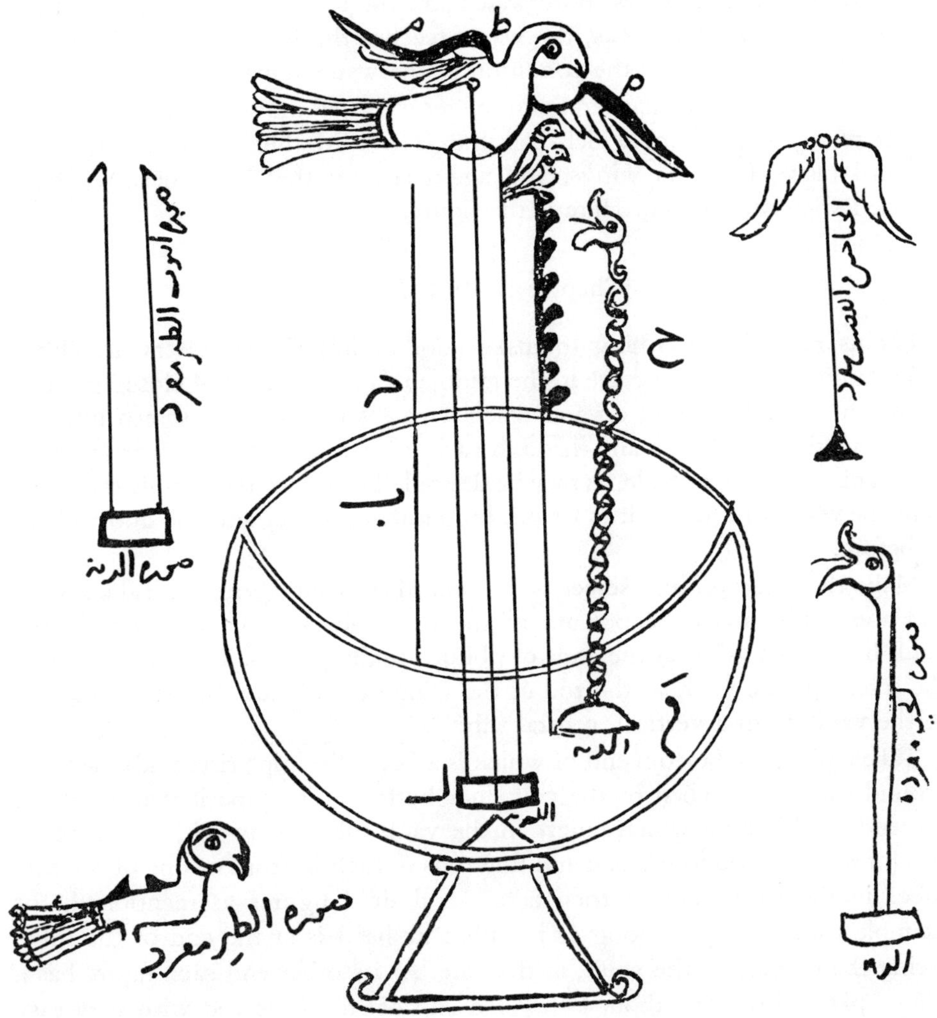

Figure 43. A.S. 3713, fo. 52b (full page). *Drawing of Chapter 40.* The Legends are undeciphered, except for "The Bird" (lower left). Reference Letters:

آ ($\bar{\alpha}$): ["Reservoir"]	ح (η): ["Snake"]
ب (β): ["Screen"]	ل (λ): ["Float (ε) and Rod"]
د (δ): ["Pipe"]	م (μ): ["Wings"]

with the snake. One pours more water and the snake rises more. When the snake is almost at the chicks, the water reaches the bird float; the bird then rises and flutters. Such is the condition as the water remains in the reservoir.

When the water is evacuated from the reservoir,[559] the snake redescends and re-enters the rock from which it came. The bird then comes back to its original place, folds its wings, and again sits on the little ones. We have finished what we wanted. Here is the figure.

[Chapter 41. Fish *Parerga*]

Let us now describe how to make another but similar reservoir with a mountain or vaulted structure in the middle of its screen and 4 closed gates in it. When water is poured on and when it passes through the screen into the reservoir, the gates open and through each of them appears a fish or snake or other animal on the outside, as may be desired. The figure stays while the water is in the reservoir. When it is let off, the figures re-enter and the doors close, as before.

Make the reservoir and screen as before. Make four openings. Take a solid rod and solder it to the bottom of the reservoir. Let the openings in the vault be window-slots at the ends of oblique [radial] tubes. The reservoir is α; the screen β; the vault γ; the rod δ; the windows ε. The tubes rise obliquely in the vault from a vertical, central tube ζ.

Take a pipe, the bottom end of which is a float. The pipe rises with the float, around the rod attached to the container bottom. The pipe is marked η. At the end of this pipe which enters in the vault-tube [ζ], to-wit, at the upper end [of η], there are four small and fine rods ϑ, each in front of one of the side tubes in the vault. Let us then take the desired figures as mentioned, for example fish. Each is as long as its tube; its head is at the end of the tube, facing the outside of the vault, at the gate. At its other end each figure has a hinge, pivoted to turn about [a hooked end of] the little rod with very easy movement. Thus, when the float rises, the figures rise and leave the doors. Each opening or tube end for one of the figures has a doorhinge. A leather trap-door is pivoted thereon by a single journal, easily turning in a fixed ring above the door.

When this is prepared as described, the water entering the reservoir lifts the float; the figures rise; they push with their heads; the doors open; the figures appear outside the doors; they remain visible while the water is in the

---

[559] It may start an intermittent siphon, of the type shown in Chapter 9 or 10.

Figure 44. A. S. 3713, fo. 54b (top). *Drawing of Chapter 41.* Reference Letters, Numerals and Legend:

آ ب ($\bar{\alpha}\beta$): ["Reservoir and Screen"]
ج (γ): ["Vault"]
د (δ): ["Mountain"]
ع (ι): ["Windows"]
ك ($\varkappa$): ["Float"]
Right and left: "The Fish"

reservoir. When the water is drawn off, the float returns to the bottom of the reservoir; the figures are pulled back; they re-enter the tubes; the doors close by themselves. This completes our description. Here is the figure.

[Chapter 42. Vulture and Sparrows as *Parerga*]

Construction of another nice vessel and very beautiful reservoir. Let another reservoir have a mountain in the middle of its screen, and on it a vulture, with lowered head and closed wings. Around him let there be sparrows with open wings. When water is poured in the reservoir, the vulture looks up and raises his wings. The sparrows fold theirs.

We make this reservoir like the first, and give it a screen, the middle of which has a hollow mountain, soldered to it all around. The chamber is α; the screen β; the mountain γ. We introduce a pipe in the hollow mountain, and make a vulture which we mount on the pipe. The bottom end of the pipe has a float. The pipe goes from the peak of the mountain to the bottom part of the reservoir. The pipe is δ; the vulture ε; the float ζ. His wings are separated like those of the bird described before; they are $\eta\eta$. The vulture's head can bend and rise, as for a cry, by inner articulation ϑ. Let a solid rod, in the hollow of the vulture's pipe, be fixed and soldered to the reservoir bottom and let it enter the vulture's body and reach to his back, as in the preceding bird construction. The two wings have rings, grouped with the ring of the head rod. The float and its pipe are so dimensioned that the vulture's talons are at the peak of the mountain, with the vulture's feet at rest, his head down, and his wings folded.

Let us then take birds, as described, and solder them only to tubes, not mountains or trees; each tube entering and reaching to a slot in the back as is done for the vulture; they are at $\varkappa\varkappa$. Rods pass through tubes in the reservoir, with floats $\lambda\lambda$ at their lower ends. The wings $\mu\mu$, with articulation, have their place at the ends of these rods. When the floats hang free in the reservoir, they pull the rods, which hold the wings, and the wings are open.

When water is poured in the reservoir, this water rises and reaches the vulture float; the float is lifted, with its pipe and with the vulture fixed to the pipe. While the vulture rises, his wing and neck articulations are pulled up; the head rises; the wings unfold. When water then reaches the sparrow floats, these floats rise, and also their rods; the wing articulations are relaxed, the wings fall and fold. This condition prevails until the water is evacuated from the reservoir. Thereupon everything returns to its place as initially. This was what we wanted to describe. Here is the figure.

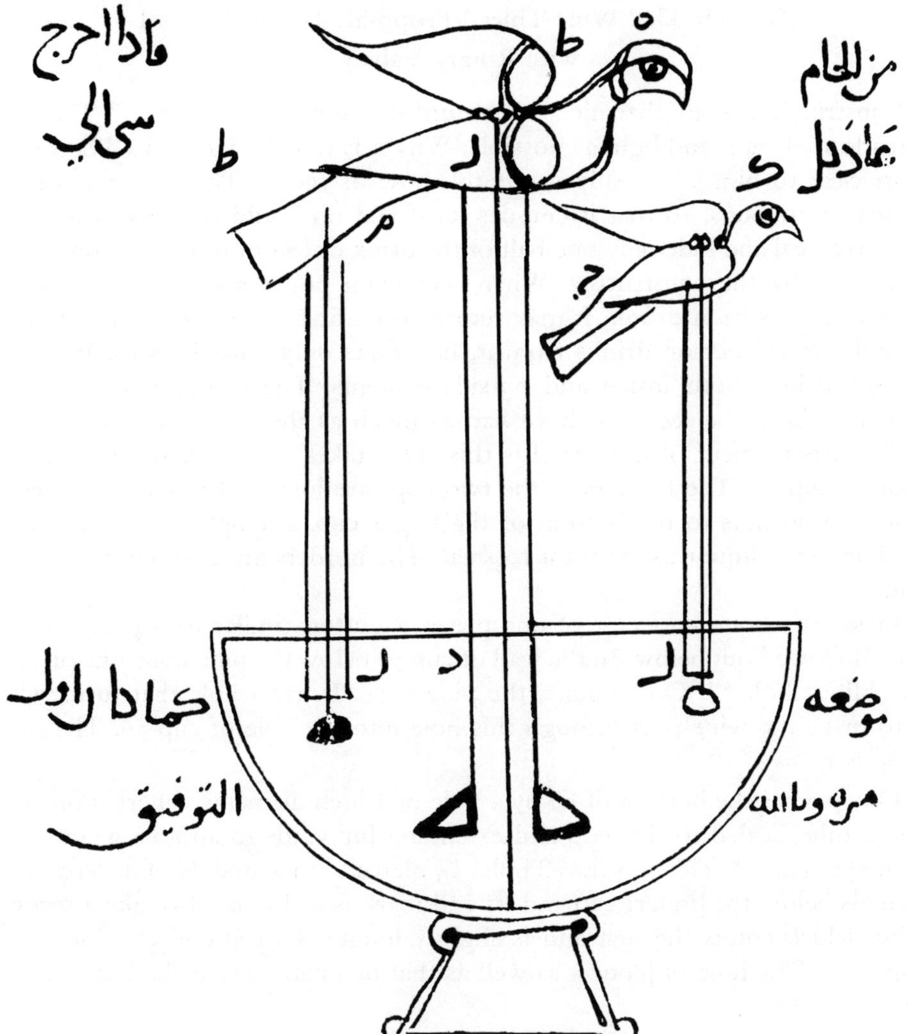

Figure 45. A. S. 3713, fo. 56a (bottom). *Drawing of Chapter 42.* The legends are poorly legible. They seem to say: "If the liquid comes out . . . as shown, for success . . . Its place . . . Allah . . ."

[Chapter 43. "Wine-Thief" Proposal: Jacketed Vessel with Rotary Valve]

Construction of another nice and beautiful vessel or "wine-thief." The cup is made as elegant and light as possible. Why it is called "wine-thief" becomes most clear to him who really wants to know its secret. This cup may drink all that it contains, so that it empties itself and no liquid remains visible, or one may steal and hide only one-half of the drink or less or more than one-half, as desired by the constructor. When one again pours wine into it in such amount as has been drunk, it may return to its first condition or may fill. If an uninformed person drinks from it, he drinks only what he sees. He who knows, drinks both inside and outside contents. The ignorant will not try to complain, as he seems to have had as much as the one who knows.

The arrangement of the vessel is this: One takes a cup $\alpha\beta$. In its cavity is another cup $\gamma\delta$. The borders of the two cups are joined. The space between, from the borders to the bottom of the larger cup, is empty.[560] We make it hold as much liquor as we want to steal. The borders are soldered to form a unit.

Inside the hollow of cup $\gamma\delta$ one places a button similar to a pine-cone, a little hollowed out below. In the wall of cup $\gamma\delta$ below the pine-cone one pierces an oblique hole.[561] One mounts the pine cone. It has a hole that turns with it to make the wine pass through this hole into the hole of cup $\gamma\delta$. The pine cone is ε.

One pierces the bottom of $\alpha\beta$ by a hole of 1 inch diameter, wherein one enters a tube, sealed to this cup and extending [upwards to an area where it is sealed] to cup $\gamma\delta$. This tube has 2 holes $\zeta\eta$ pierced in the middle of its length. It extends below the [inner] cup and its end mark is ϑ. Let us also take a second tube, which enters the first and is slightly longer. Its top end is soldered to cup $\gamma\delta$.[562] The hole of [cone ε as well as that of] cup $\gamma\delta$ is at the head of this tube.

When wine has been poured in, [filling the jacket, and when the holes coincide,] it enters through [the holes of the 2 tubes and] the hole of the pine cone in the hole of cup $\gamma\delta$, the [inner] tube having 2 holes pierced in it opposite

---

560 In this space, the drawing shows odd lines. Perhaps they originally were construction lines, to be erased finally, which the draftsman, in the end, erroneously executed with ink.

561 So the text. In fact, a ring-slot seems to be meant. Also see Chapter 46.

562 As noted by Carra de Vaux, the "pine cone" must be meant. The protruding bottom end of this tube may be intended as a bearing.

those of the first tube; they are at μ. The inner, turnable tube is closed by a wall, below holes μ. It is made tight against leaks anywhere, but is well fitted into the tube soldered to the [larger] cup, to turn in it with friction. It lets no liquid pass [except through μ]. On the lower part of the turning tube one fastens a disc,[563] as support for the [outer] cup soldered to the first tube [and one thereby prevents the cone and inner tube from being lifted out]. Whoever sees

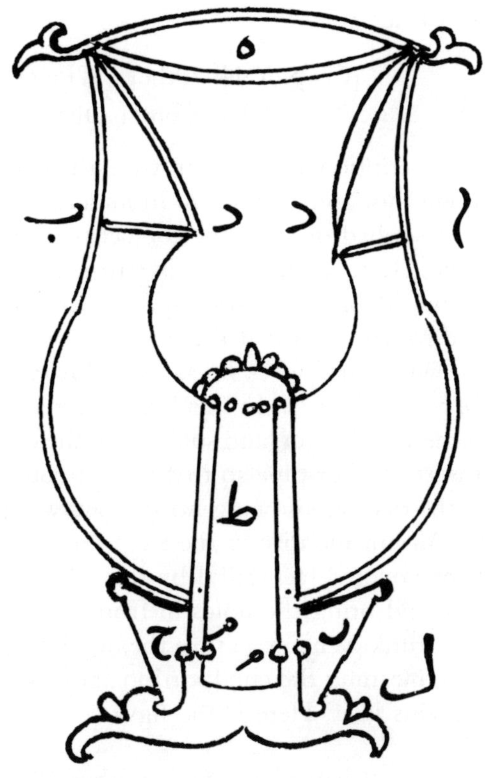

Figure 46. A. S. 3713, fo. 58a (middle). *Drawing of Chapter 43.* Reference Letters:

ا ب ($\alpha\beta$): ["Cup"]
د د ($\delta\delta$): ["Inner Cup ($\gamma\delta$)"]
ط (ϑ): ["Inner Tube, soldered to inner cup"]
ل (λ): ["Disk"]

the cup, believes it is a single cup and does not suspect what we have placed in it. The disc is λ. So far the description.

A marker is scribed, outside the tube [and disc structure partly] soldered to the cups, on the back of the larger cup $\alpha\beta$, and another marker on the disc. Each marker can be a thin line or some other recognizable sign. This marker [on the disc] is in front of the holes [of the inner tube], which are [when the

[563] The drawing is wrong as to the location of the holes and as to details of the disc.

valve is open,] the holes through both tubes. When they are open, wine poured in will enter and fill the space between the two cups, and then the cup $\gamma\delta$. If one leaves the holes open as they are, and drinks the contents of cup $\gamma\delta$, the wine in the space between the cups also follows. If one closes the two holes by turning the disc so that the outer marks no longer coincide, one drinks only the contents of cup $\gamma\delta$. What is between the cups then remains locked-in. The earlier part of our book has explained why this is so.[564] We need not repeat it.

[Chapter 44. Other "Wine-Thief" Proposal: Vessel with Horizontal Divider and Rotary Valve]

Description of another wine-thief vessel. Take chamber $\alpha\beta$ and divide it along [horizontal] line $\gamma\delta$. In its middle pierce a hole ε; on it place a pine-cone as described or any other object to conceal the hole. Construction of the pine cone: Take a button similar to a pine cone, hollow inside. Let the hole be below it, in the cut-out space. Let the cone be pierced by a concealed hole and mount it so that the wine passes through this hole and nowhere else. Let the wine-thief space be at $\zeta\eta$ [below the partition]. Pierce the bottom wall of $\zeta\eta$ by a hole ϑ.[565] In this hole insert a pipe as long as the stem of the vessel or somewhat longer, and solder it to the cup. At a point of this pipe pierce a hole ι. Insert another pipe so that it is surrounded by the first and turns in it exactly, with friction, allowing no leak anywhere. Pierce it by hole $\varkappa$ in front of hole [ι]. Mount the disc of the stem on the bottom end of this [inner] tube and let it be covered [centrally] by the tube soldered to the cup. When you turn the disc and bring the holes in front of one another, the full contents of the cup are drunk. When you turn the disc back, only that is drunk which is seen in the cup, meaning the cup [portion above the partition] $\gamma\delta$; what is in the reservoir remains in it. Here is the figure.

[Chapter 45. Still another "Wine-Thief" Proposal: Jacketed Vessel with Movable Divider]

If you wish, you can give another form to such a vessel, and make a cup which rises and sinks. Take a cup of the desired form. This is the large cup $\alpha\beta$. It has a wall which is hollow as shown, from α to γ and β to δ. In the hollow

[564] It is not clear what part is meant. The book rather explains that a construction as shown and described here is *inoperable*, in the absence of a vent for the jacket. Perhaps Chapter 46 is meant; that is, this copy may show the "earlier part of our book" *after* the present part.

[565] From here, the drawing is unfinished, as is that of Chapter 45.

space of cup $\alpha\beta$, a cup ε is fitted. The bottom of cup $\alpha\beta$ has a 1-inch hole. Cup ε has a long stem $\zeta\eta$ soldered to it. When it is raised, cup ε rises to line $\gamma\delta$. Make a mark $\vartheta\iota$ at the end of the stem of cup ε, and a mark $\varkappa$ below cup $\alpha\beta$; the distance between these marks equals the distance from ε to $\gamma\delta$.

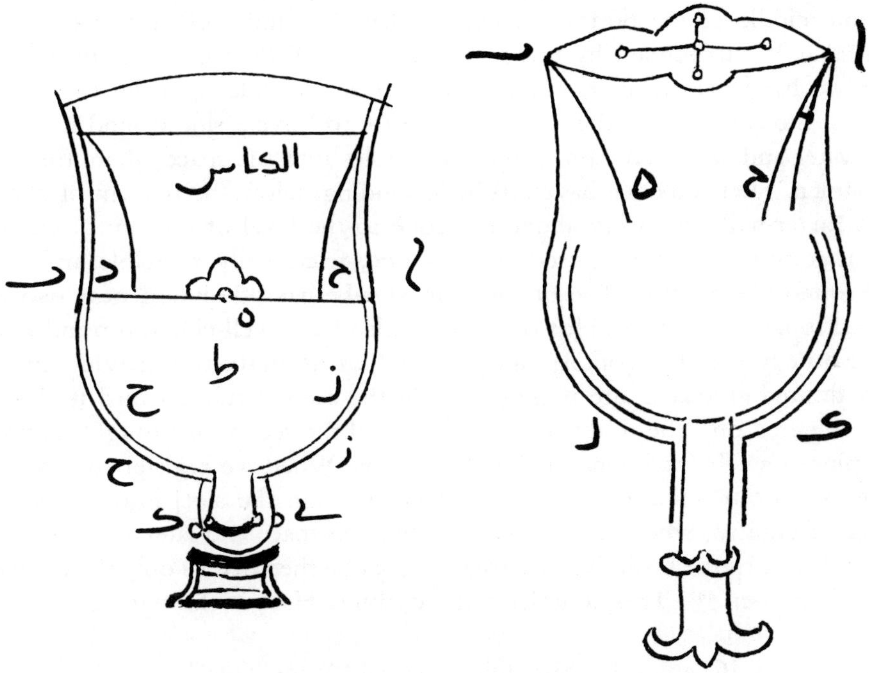

Figure 47. A.S. 3713, fo. 59a (top) and b (bottom). *Drawings of Chapters 44 and 45.* Reference Letters: similar to Figure 46.

The cup-bearer holds the cup. If he shall serve a little, he raises the stem to the point where the mark on ε shows contact with $\gamma\delta$. He then pours wine; it only fills cup $\gamma\delta$. If he shall serve much, he pulls the disc down. Its cup descends, and the wine then fills vessel ε and vessel $\gamma\delta$.[566] This completes the chapter. Here is the figure.

566 This latter point—the only point of the chapter—may be true or untrue. Depending on the interpretation of the unfinished, schematic drawing, wine may also be drawn into the lower, outer chamber by *raising* the disk in the "fitted cup" to rarefy the air in this outer chamber. A very unsatisfactory copy of a chapter with technical refinement, naively recorded.

[Chapter 46. Vented "Wine-Thief"]

Construction of another wine-thief vessel. If you wish, make the vessel as follows. Make common cup $\alpha\beta$. Insert another cup $\gamma\delta$ in it. Let the two rims meet exactly and be soldered together. Between them is a void for the wine. In the middle of the bottom of cup $\gamma\delta$ drill two holes of proper size, or a central hole surrounded by three or four holes. Take one of the pine cones that we have described, having centrally a single hole opposite the central hole of the cup. Mount the cone, taking care to leave a slot around it so that the water and wine can pass through the holes into the space where the wine is "stolen." Let the cone have a tube, extending below the bottom of cup $\gamma\delta$ by a finger-nail and rising above the cone to the level of the rim of cup $\alpha\beta$. Another tube is superimposed on this one, frictionally rotatable and tight all around like a valve. The first [or inner] tube has side hole ζ near its well-closed top, and the second [or outer] tube also has a well-closed top and a hole η opposite hole ζ. This [outer] tube [as noted] turns on the tube [having hole] ζ. Any desired animal figure is mounted on the second tube; a bird is shown. It is soldered on. The tubes are assembled and the second is brought down to the pine-cone. It then conceals the inner tube. When we register the the two holes and pour wine in cup $\alpha\beta$, it enters [through the slot] into the hollow space of tube $\gamma\delta$, where it is "stolen" It fills the space and later returns to $\alpha\beta$. If we then turn the bird, the two holes close; one then drinks only the contents of a single cup.[567] This completes the chapter. Here is the figure.

[Chapter 47. Simplified Vented "Wine-Thief"][568]

Construction of another wine-thief vessel, more beautiful than the preceding one. This vessel can also be arranged in another way. Make a vessel in a vessel. The bigger one is $\alpha\beta$. Its inside contains another, $\gamma\delta$. Their rims meet and are soldered together. The space between them is the place where wine is "stolen." In the middle of the bottom of vessel $\gamma\delta$ drill 2 or 3 small holes ε. Superimpose on these holes the figure of some animal, as desired; let it cover its litter with its belly, the feet being near holes ε. Solder it on by its feet, letting its belly conceal holes ε. Provide a hollow handle for the vessel, with a hole which opens

[567] "Because, as the two no longer allow passage of air, the liquid between the two cups is imprisoned" (Carra de Vaux, footnote). Chapters 46–50 are based on archaic vases of a type shown by Deonna, Fig. 2. These are operable, in contrast to those of Chapters 43–45.

[568] Drachmann considers the cups of Chapters 47–50 "original," that is, Philonian. See note 569.

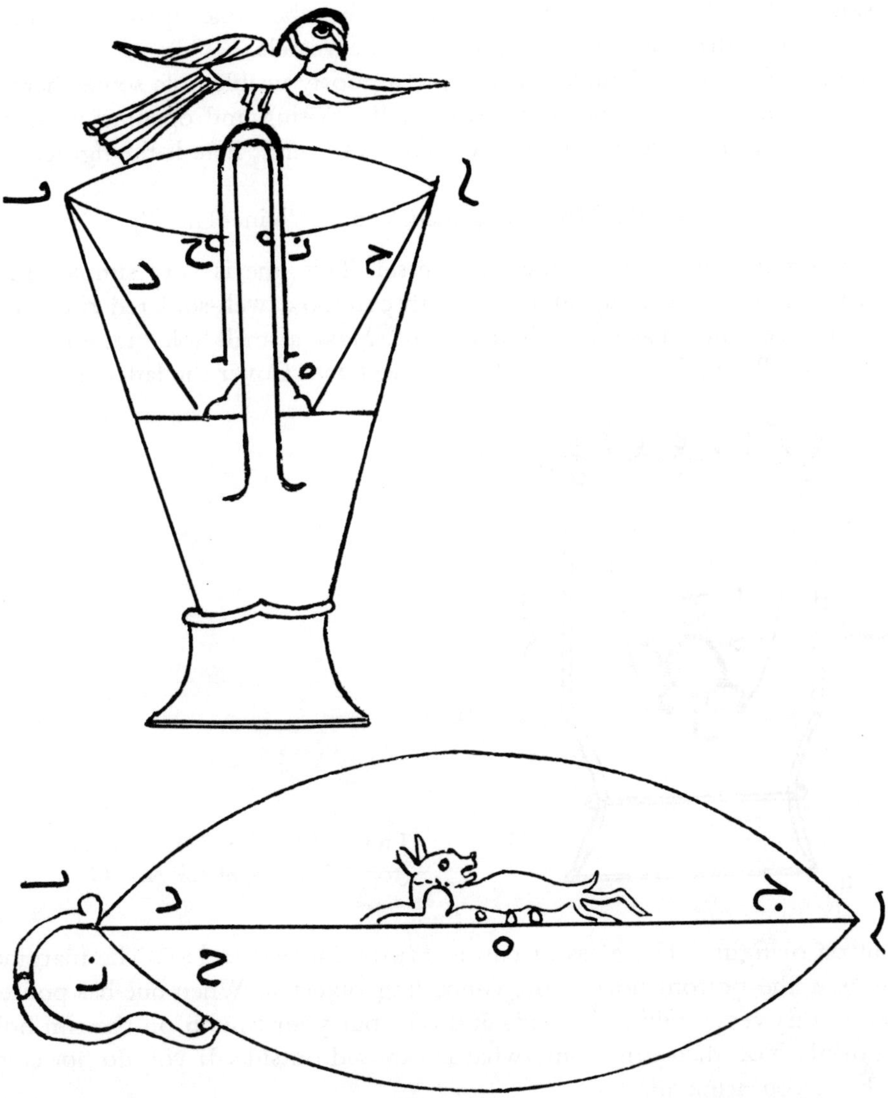

Figure 48. A. S. 3713, fo. 61 a (top) and b (bottom). *Drawings of Chapters 46 and 47.*

into the hollow "thief" space between the vessels. In the middle of the handle there is a hole ζ; the hole to the chamber is η. When wine is poured into this vessel, it fills the "thief" space as well as the open space, but if one then closes the handle hole with a finger and drinks, one drinks only the liquid that appears outside—one-half of the entire contents. Only when one opens the handle hole does one drink all of the contents of the vessels.

Instead of using the handle one can put a poorly visible hole somewhere in the vessel [as shown in the next chapter]. By closing and opening it with a finger one can produce the same effect. This is the end; here is the figure.

[Chapter 48. More Simplified Vented "wine-Thief"]

Construction of another wine-thief vessel. This one is very simple. Take a glass in a glass, or a cup in a cup, with common, well-soldered rims. The space between them has the desired volume. Make a small hole in the rim and another small hole in the middle of the [inner] cup. Cover the latter hole with

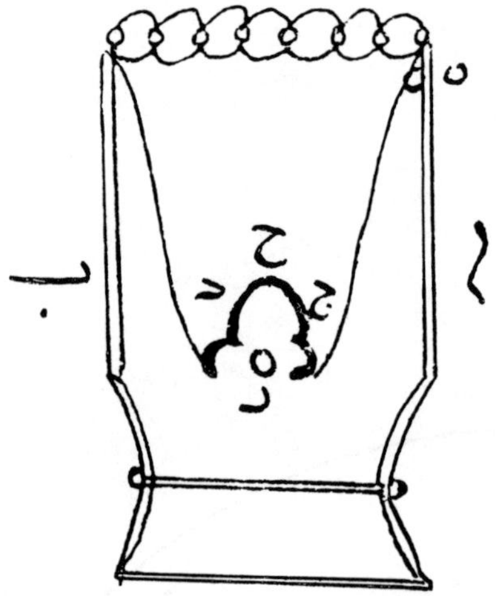

Figure 49. A.S. 3713, fo. 62b (top). *Drawing of Chapter 48.*

a button or figure. The glass or cup is $\alpha\beta$; the inner vessel $\gamma\delta$. The marginal hole is ε, the bottom hole ζ, the concealing object η. When one has poured wine in this vessel, filling it inside and out, put your thumb on the rim hole and drink. You then drink only what is exposed outside. If you do not close the hole, you drink all.

Let this be understood, and notice that it can be varied at will. The operation resides entirely in the hole and its closing and opening. This is the end of what we wanted to say about this kind of devices. Here is the figure.

[Chapter 49. Jacketed Constant Level Cup, with Siphon]

Construction of another vessel, for "balanced" operation. This vessel can always have only the same amount of liquid poured into it. If one pours more, everything runs out. You can make this vessel like a drinking cup or glass. It is marked α. Inside is another vessel, attached to it, with little space between; this is $\gamma\delta$. The drinking glass holds more than a *ritl* if one mixes water with wine, or it holds exactly a *ritl.* Place a bent siphon in the space between the two vessels, with an inlet opening on the bottom of glass $\gamma\delta$ and rising to the level of the straight line $\varepsilon\zeta$. Bend it so that its other end comes out at the bottom of the vessel at η; let the inner opening be at ϑ. The vessel has 1 or 2 handles if we wish. The siphon is not seen, being entirely in the space between the 2 vessels.

If we complete this construction and if we then pour wine, mixed with water, up to the straight line $\varepsilon\zeta$, the wine [and water] remains and does not run out, but if we overstep this line, all runs out at the bottom of the vase at η. We have explained it and given the cause in the preceding account. Here is the figure.

[Chapter 50. Jacketed Cup, for Drinking without Tilting]

Construction of another vessel, a surprising drinking glass. One drinks from it without tilting it. Take the glass $\alpha\beta$ which has another glass in it,

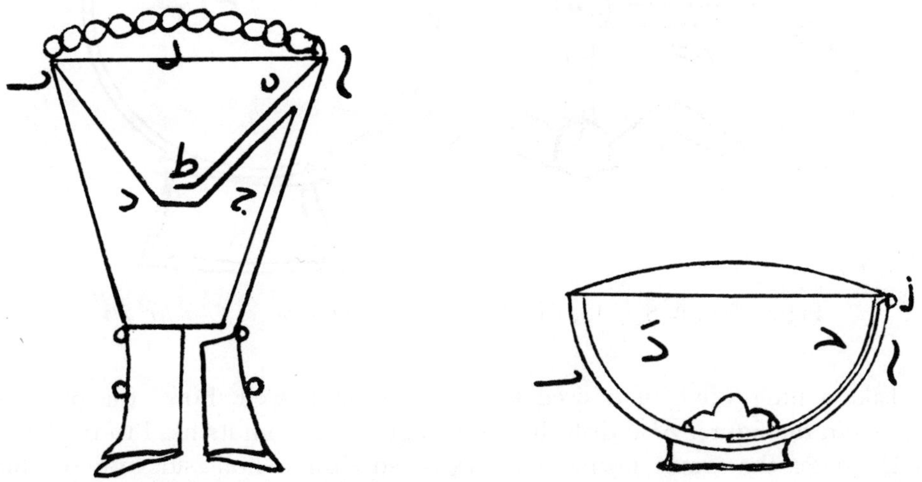

Figure 50. A. S. 3713, fo. 63a (bottom) and b (bottom). *Drawings of Chapters 49 and 50.*

their rims being united and the second being $\gamma\delta$. Not much space is needed between them. Provide a bent pipe between them, with an opening in the middle of glass $\gamma\delta$ and another opening on the rim of the glass, at ζ. Mount any desired figure on opening ε, letting wine enter below it and concealing the opening. When the vessel is full, the user takes it, puts his lips on opening ζ and sucks out the contents. He can thus drink all of the liquor without tilting the vessel. Let our description be understood. Here is the figure.

[Chapter 51. Faucet Combination I][569]

Construction of another vessel: an elegant and surprising faucet. With this faucet, the water comes out through the principal outlet or through [one in its] handle, as you wish.[570]

Figure 51. A.S. 3713, fo. 64b (bottom). *Drawing of Chapter 51.*

Take a plug valve of desired kind and finish it well. Form the plug as shown in the figure. Also drill the plug lengthwise, from its head to the large hole where the water normally flows. Also drill a hole sidewise of this principal hole, and exactly into the lengthwise hole. By turning the plug you can close the principal water outlet on the side of the reservoir.

---

[569] Ground stop-cocks and the like were described, in clumsy terms, in Chapters 22 and 43–46. Here, by contrast, ground stop-cocks are taken for granted and are developed further. Also see note 478.

[570] A dotted line is used to indicate the latter alternate: a graphic technique rarely used until well after the middle ages. Ibn Musa, Fig. 22, uses a similar line to indicate a valve-actuating chain.

Take the handle part [or head, drilled lengthwise] and drill a hole into the [lengthwise] water-passage, in the handle-ornament. Close it by a little plug valve when the main passage is open; open it only to let water come out through the handle.

To close [both] the main outlet and the handle [outlet], turn the [main] plug left; this closes both passages. To draw water only through the handle, turn it right. To draw it from the main outlet, turn it into the direction of the main outlet, and it then comes out only there. The principal hole in the plug is ϑ. The inlet to the handle is ε; it is drilled sidewise. The hole through the handle is $\varepsilon\delta$. Let it be understood and make it well. Here is the figure.[571]

[Chapter 52. Faucet Combination II]

Construction of another vessel and faucet, more beautiful than the preceding one. It is for a partitioned vessel, where the water or wine comes out at the

Figure 52. A. S. 3713, fo. 65b (top). *Drawing of Chapter 52.*

571 The auxiliary faucet may serve as a sample outlet. Carra de Vaux shows it in upwardly squirting form; he gives no reasons.

same point. Take the vessel, divided into two halves lengthwise by a partition. Take two thin tubes $\alpha\beta$, carefully soldered together with lead, and put them into the hollow of the large valve pipe, which serves as female member for the principal water outlet. They end at points $\gamma\delta$ on the valve plug, which points correspond to the thin tubes. They, that is, $\gamma\delta$ in the plug, open to the hole ε on the side of the main outlet, to discharge water and wine. When you turn the handle right, hole γ is closed, hole δ stays open; an outlet is provided by common hole ε. When you turn the handle left, δ closes, γ opens, ε stays open. When you turn the handle into the direction of the main outlet, both holes close and nothing flows. Let it be well understood and well made. Here is the figure.

[Chapter 53. Faucet Combination III]

Construction of another elegant and surprising faucet. We take an ordinary faucet, as desired, according to the figure. The large ornamented stem with

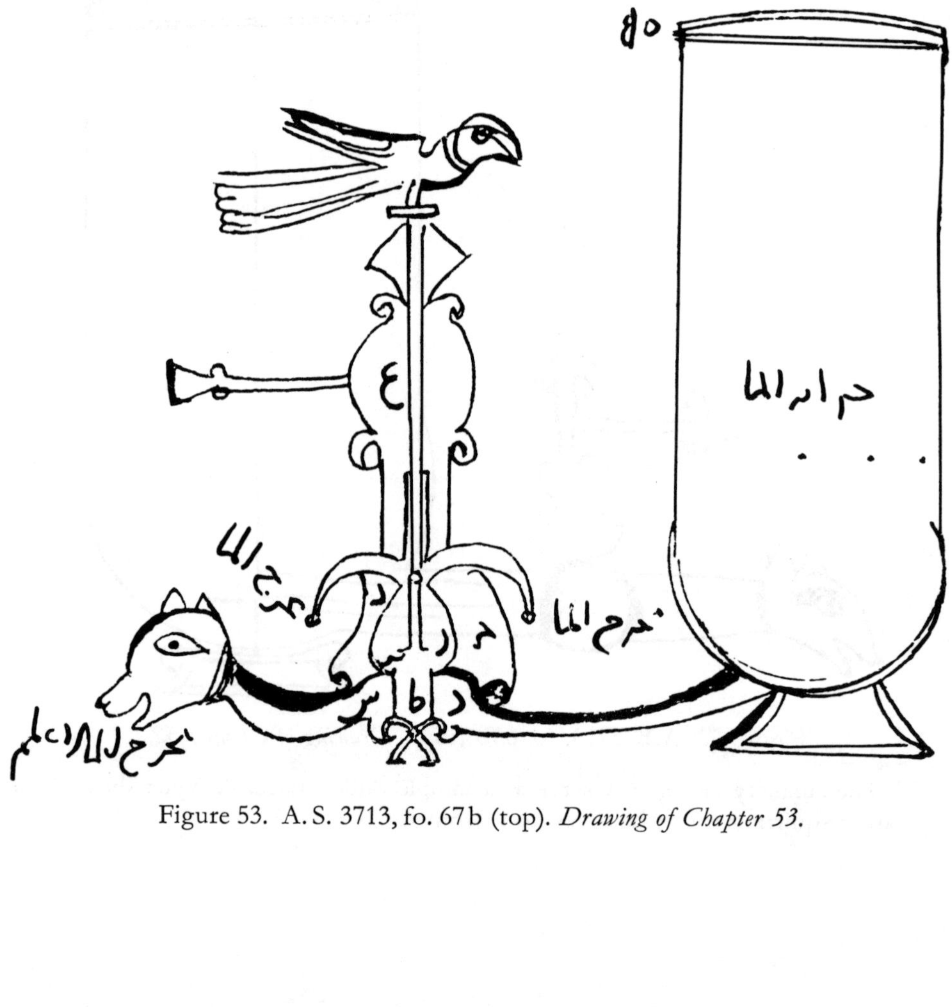

Figure 53. A. S. 3713, fo. 67b (top). *Drawing of Chapter 53.*

outlet trunks for water and wine is fixed and immobile. It has four hollow nipples,[572] disposed around it for alternatingly discharging water and other liquid as one turns.

The [next inner] ornamented stem, which supports the handle of the plug valve, is $\vartheta\eta$. The main water outlet hole is ε. There is a long hole ζ in the side[?] cavity of the valve, longitudinally drilled as described above and with connections to the nipples $\gamma\delta$. Male member $\upsilon\mu$ has a bird on its head. It turns easily, when desired. This male member continues into the plug handle $\eta\vartheta$, somewhat below the nipples. It has orifice σ which can be registered with any one of the nipples. The handle is well made and finished so that, depending on its turning relative to the main outlet, liquid is discharged, or if it turns right or left, nothing is discharged. As one turns the bird and as the handle passage then registers with any of the nipples, liquid comes from that nipple. Let this be understood. Here is the figure.

[Chapter 54. Faucet Combination IV]

Construction of another elegant faucet, with water wheel. Water is discharged from the handle or head of the ornamented plug, through a jet,

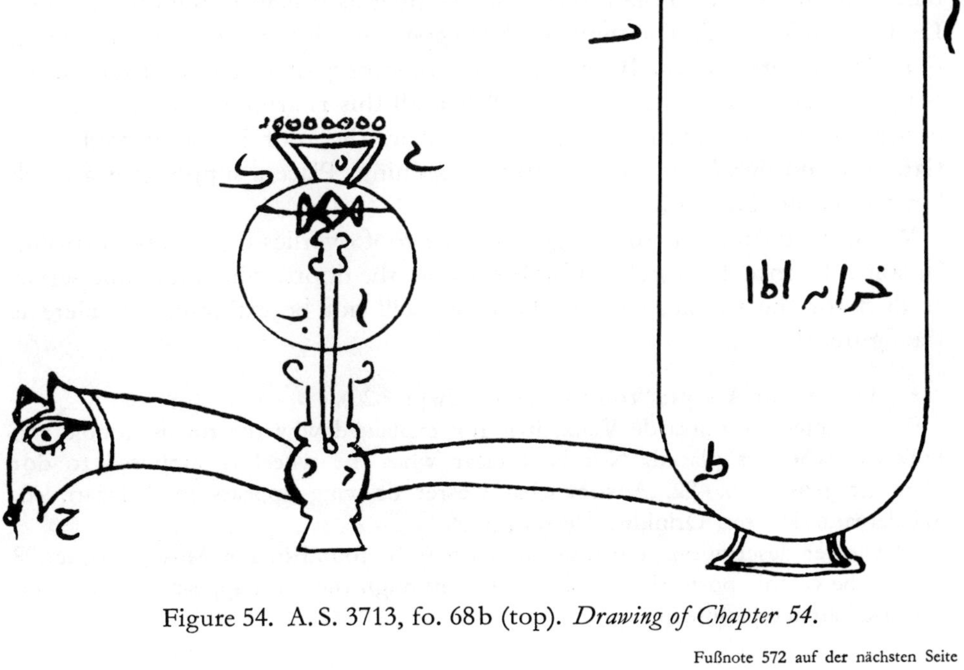

Figure 54. A. S. 3713, fo. 68b (top). *Drawing of Chapter 54.*

Fußnote 572 auf der nächsten Seite

whereon a turning mill-wheel is interposed. We drill the principal water inlet and outlet and mount the valve plug in the faucet. The plug is constructed as follows. Let there be two hollow rods $\alpha\beta$ having apertures $\gamma\delta$ in the plug. A hole ε opens to the length of the pipe, where the wheel turns. The water from the two rods is directed onto vanes $\iota\varkappa$. It then leaves through small screen apertures. The water enters the two rods through a single hole η in the valve plug on the side of the main water inlet, as arranged and described above. When the handle is placed in the direction of the main outlet, the water discharges through this outlet. When we turn the handle, the wheel begins to turn. The vanes and all their parts are constructed for easy turning. If desired, the handle can have a small plug for closing and opening it; thus it becomes more convenient. Let it be understood and well executed.[573] Here is the figure.

[Chapter 55. Floating Lamp]

Construction of a wooden chamber with four open doors, and a lighted torch in it. It can be submerged in a water current and placed on its bottom, where it can be left as long as one wishes. When one retrieves it the torch is still lighted and not extinguished. If something is in the chamber, it is not wetted.

Take an ordinary chamber with a bottom which can be raised and lowered. The four corners serve as guides for this mobile boarding. This can be raised, all the way into the roof of the chamber as if it were the ceiling itself. Let it be understood. The framework is $\alpha\beta\gamma\delta$. The closure or mobile boarding is ε. The doors are $\zeta\eta$. If the chamber is strong, it can have four doors. Supports for the chamber are at $\vartheta\iota$. When all this is arranged, take a pipe of desired length, connected with the chamber, the roof being pierced at σ. Carefully and lightly fit the tube to this opening. Place a support for a torch [on the mobile bottom].

When the chamber is submerged and the roof overlies the mobile boarding [at some distance, the torch is visible through the doors, while] the tube serves as duct for the smoke, so that the torch will not be extinguished. Here is the figure.[574]

[572] The text in **A** says: "trunks" (as in Chapter 20).

[573] As noted by Carra de Vaux, it is not explained why the rod is double and how the wheel is placed. Nor is it clear what the wheel is supposed to do: Hammer-Jensen, p. 482. A somewhat clearer drawing appears in AlJazari, ed. Wiedemann-Hauser, Drinking Vessels, p. 59.

[574] Clearer descriptions and illustrations may be found in Ibn Musa, Chapter 98 and in the Gothic postscripts of **C** and **H,** although they still appear to depend on confused archetype information.

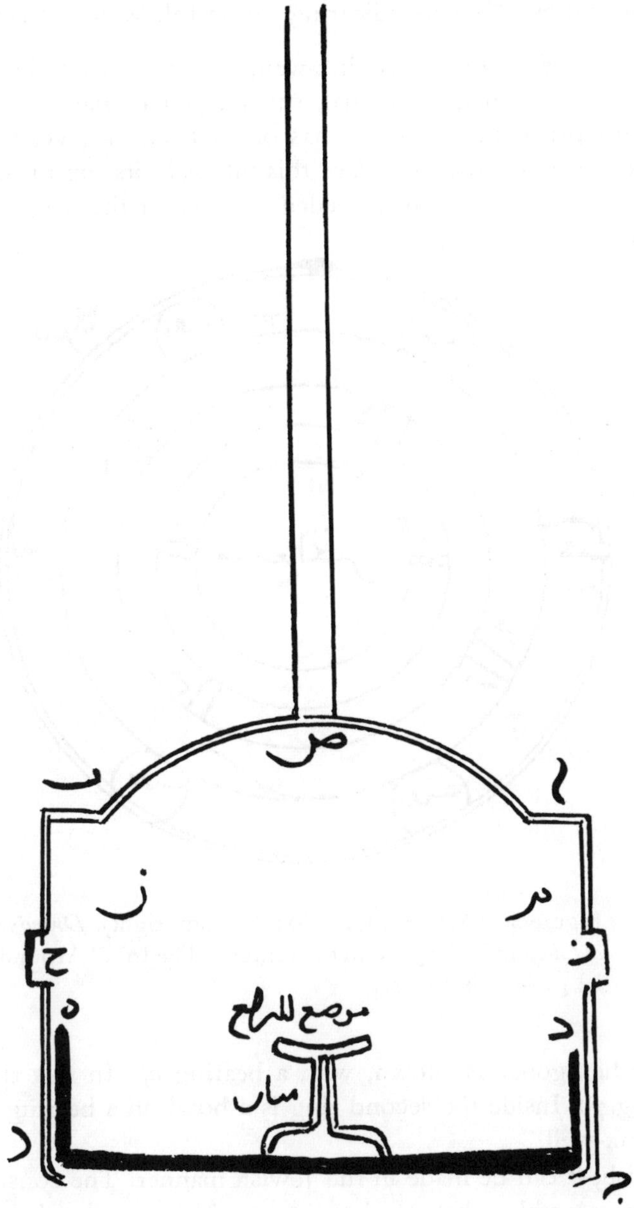

Figure 55. A. S. 3713, fo. 69a (full page, with two columns of text, right and left of the stack). *Drawing of Chapter 55.*

[Chapter 56. Universal Bearing for an Ink-Well or Censer][575]

Construction of an octagonal ink-well, a very elegant device. We make an ink-well of octagonal, hexagonal, square, pentagonal or other form, any form used for prismatic glasses. It has on each face a place from where one can write. Regardless how you place this ink-well, its top presents a hole for entering the pen. No reversal is needed. You enter the pen, it finds the ink, and you write.

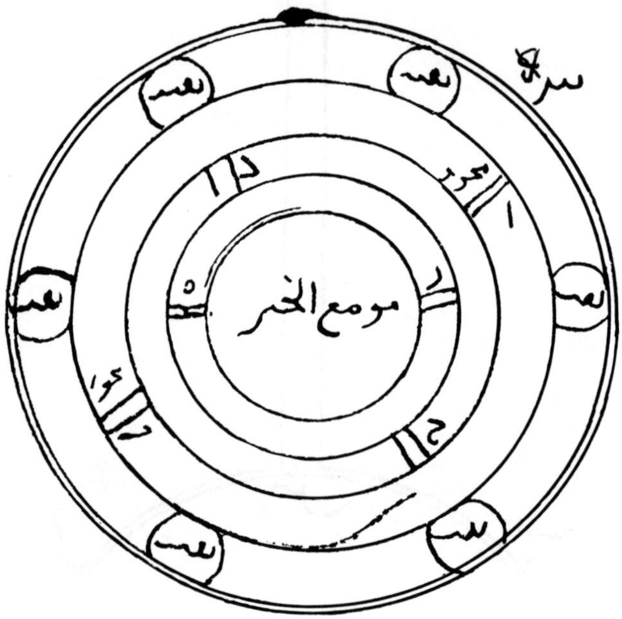

Figure 56. A. S. 3713, fo. 70a (bottom, right). *Drawing of Chapter 56.* Legend in the center: "The Ink." Around it: س (σ), six times.

Let it be hexagonal as shown, with a bearing $\alpha\beta$. In this ring is another, on a bearing $\gamma\delta$. Inside the second ring is a bowl on a bearing $\varepsilon\zeta$. This bowl forms the ink-well.

If you wish, it can be made in the Jewish manner. The construction of the device then resembles that of the censer which remains balanced while it turns.

---

575 About Gothic parallels to this chapter on a "Cardanic" suspension, see the separate note, above. Also see Beck, p. 71; Drachmann, p. 67.

Measure it well and adjust it carefully so that, regardless of the face on which the ink-well rests, you face the top of the ink-well when entering the pen. Here is the figure.

[Chapter 57. Perfume Burner with Steam Boiler for Self-Kindling][576]

Construction of a perfume-burner which fans itself. Take a perfume-burner which has a water-reservoir vessel below its grate. The burner is $\alpha\beta$; the place into which water is poured is $\gamma\delta$. From [above] the place of the water departs

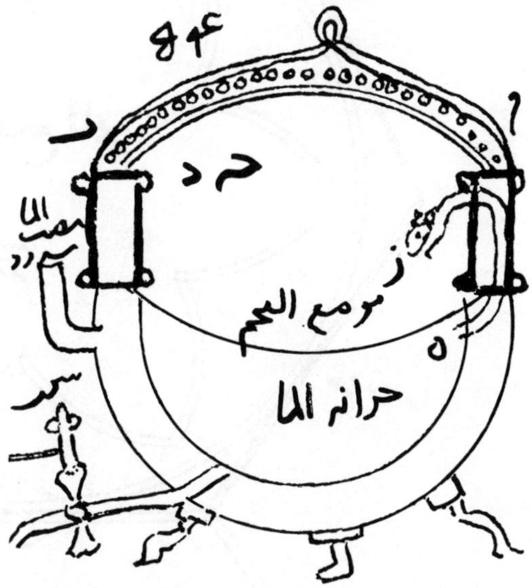

Figure 57. A. S. 3713, fo. 70b (bottom, left). *Drawing of Chapter 57.* Legend in upper compartment: "Coal." In lower compartment: "Water."

a curved pipe $\varepsilon\zeta$, which goes to above the grate. Otherwise everything is built as in perfume burners generally known. When the water boils, it vaporizes. The vapor issues from the end of the curved pipe, above the fire, which it fans. A water outlet is provided by faucet η. Let it be understood. Here is the figure.

[Chapter 58. Whistling Tower][577]

Construction of a Whistling Tower. Such towers can whistle as steam issues from the beak of a bird or through the mouth of a figure with breathing

576 Again, see the parallel, Gothic fragments.

577 Chapters 58–63 are an ill-assorted compilation of pneumatic-acoustic devices. In principle, such devices had been used since prehistoric times. See for example:

chest. The tower is $\alpha\beta$, the water reservoir γ, the grate δ, the water inlet ε—a funnel with valve—, and the water outlet ζ. The tower has [vertical] pipes around it [like one of the perfume sprayers], coming from outer parts of [the steam space, provided here above] the water. Mount any desired figures on

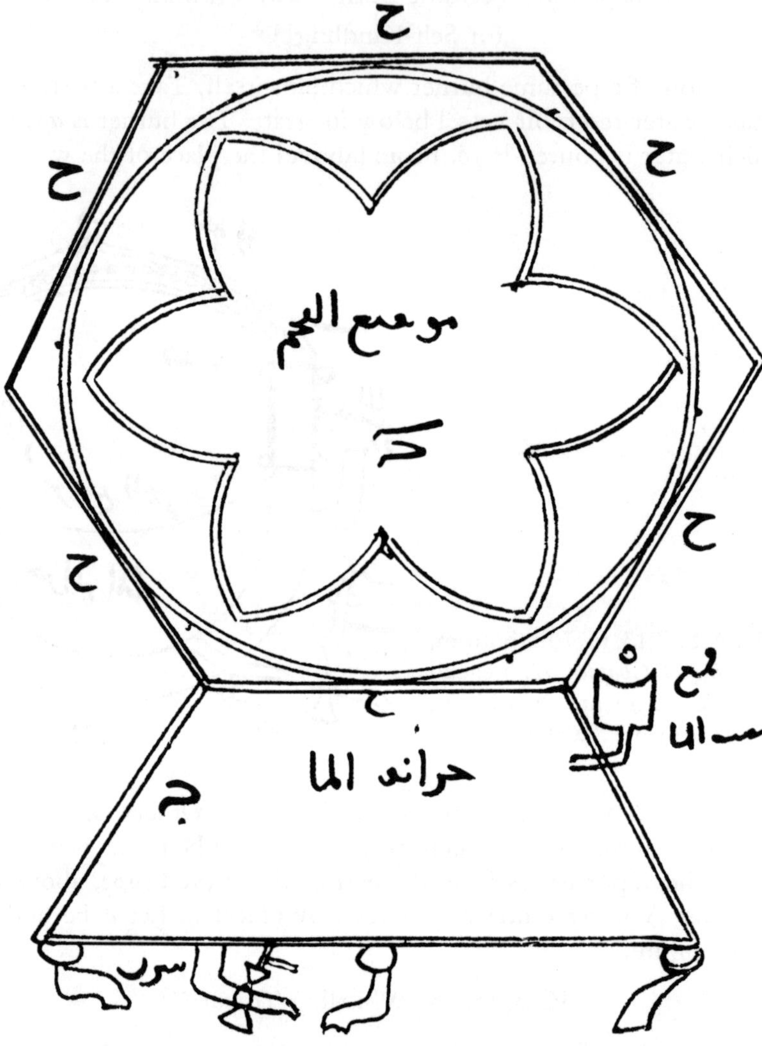

Figure 58. A. S. 3713, fo. 71 a (top). *Drawing of Chapter 58.*

Dictionnaire archéologique des techniques, Paris, 1963–64, I, p. 231: a whistling vase from prehistoric Peru. The scientific study of pneumatic acoustics may have been begun by Ctesibius, according to the remarks of Vitruvius.

them, as described: birds or others. When the water boils the steam rises through these pipes and issues where sound shall be produced. Black rings on the necks of the birds mark the whistles. This is the description of this very beautiful device. You can execute it as you wish. Here is the figure.

[Chapter 59. "Dragon and Pan Automat" with Interacting Siphons and a Variable Bypass Valve or "Cup"]

Construction of a device: a vessel for use in a temple. It must be near a source or a running water coming from a cave or cliff. It is placed in a safe place such as a temple [-court]. The water should drip from some orifice or crevice. Near it there is a dragon or a stag or whatever figure fits the temple or other place where the device is to be located. The figure stoops to the waterfall as if to drink. Before it is a person, called Paniscus or Hermarion in Greek, in a posture as if to restrain the animal's drinking. He stands on a cover and can be moved and returned to different positions. In order that the dragon should drink, this Little Pan must be turned [to face] away. Meanwhile the water runs and the dragon drinks and slurps with great huffing and puffing, as if it was very excited. If you wish you can put an elegant basin under the small waterfall.[578] The animal then drinks all the water that flows into it. However, when the Little Pan is turned to face it, the animal ceases drinking as if the figure restrained it. The device is constructed as follows.

One makes a copper container of 10 *kail* capacity. Tube β enters its bottom on one side and [curved siphon] tube γ on the other side. The entire device is placed in a dugout in the ground and is covered with a cover-sheet arranged to simulate a rock so that nothing [of the container] appears. The curved tube γ, which is [also] invisible, rises to the level $\delta\varepsilon$.[579] The [visible] ground line is $\zeta\eta$. The dragon is ϑ; the Little Pan ι; and the water outlet [from the visible basin to the concealed container] $\varkappa$. The dragon has a pipe $\lambda\mu$, which extends from its mouth through its neck and bifurcates in its two legs. It then enters the container downwardly. All this is well sealed and strongly soldered everywhere, as is necessary for the success of any pneumatic apparatus; if there were a leak anywhere, it would not operate.

---

578 This basin, or in its absence a simulated natural pool of water, could be made deep enough to start the dragon-siphon by immersion.

579 That is: the top of this tube establishes the siphon's starting level. The *kail*, a measure used only here, had 6 to 22 liters, at different places and times: Hinz[103], p. 40. The development seems related to Ibn Musa, Figures 5 and 6, but the surviving copies of the Ibn Musa descriptions are too corrupt to allow definite conclusions.

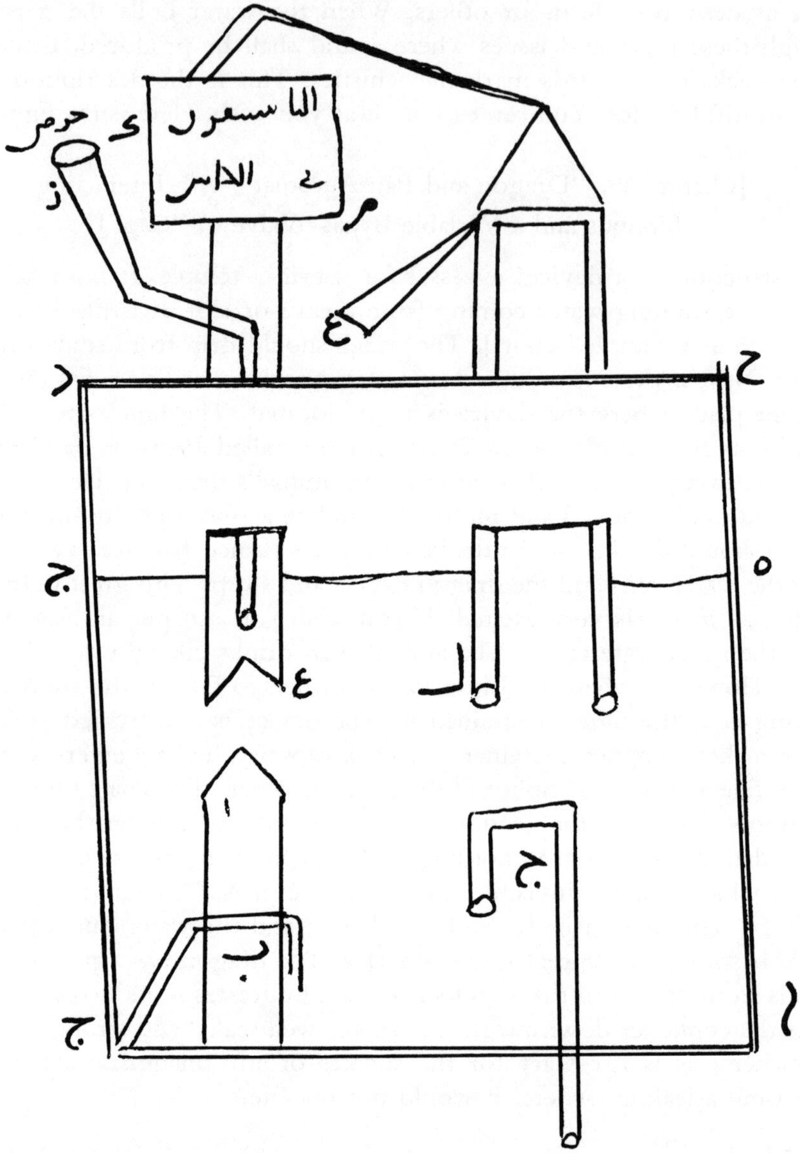

Figure 59. A.S. 3713, fo. 73b (full page). *Drawing of Chapter 59.* Legend in the rectangle, upper left: "Bowl ... for the Siphon."

A small, elegant pipe must bring water to the basin [above the outlet].[580] When this basin is full, the curved tube [of the dragon] also fills; water overflows and leaves through this curved siphon tube. The said pipe brings more water into the basin than the basin discharges, so that the basin [normally] is [more or less] full. An amount of water discharges downwards through the cover-sheet of the container to a tube ν.

Paniscus stands on a copper shaft, which goes down to the bottom of the container. Below him, [in the container,] a small tube is soldered to [and around] this shaft. It has at its [upper] end [a side tube carrying] a cup or the like. This tube [and side tube unit] is $\upsilon\nu$; the cup is σ; the shaft of the Little Pan supports part or tube υ. When Paniscus is turned to face the dragon, [cup σ is removed from below $\varkappa$; the basin in front of the dragon drains, and] the dragon seems to stop drinking [while the lower container, invisibly, fills through $\varkappa$]. When Paniscus is turned in the other direction, the cup comes below the inlet tube [$\varkappa$] of the container, and leads the received water to point β and away from the group.

When water thus no longer flows into the container [at $\varkappa$ but through the dragon siphon], the container is evacuated [by its other siphon γ. Refilling of the container with air is impeded: $\varkappa\sigma$]. The container can then only receive air, together with water, through the dragon's mouth, both being aspirated with great force. When water [from the source then] flows rapidly, he drinks greedily. When it flows slowly, he drinks in proportion, for with the cup under the outlet, he drinks in large amount so long as [the container is partly void and accordingly] the aspiration continues. When the cup is returned, he stops drinking; it seems he no longer dares to drink. This is the arrangement of this beautiful device. Here is the figure.

[Chapter 60. Whistling Bird Automat]

Construction of another really nice apparatus of the same type as the first.[581] We shall build another, similar device, operated by water from a cave or source or other origin of its flow. Let small, well-shaped birds be made, sitting on small rocks or small trees. Let them have different whistles in their throats, to sing different sounds. Let them sing while the water runs, and

---

[580] The scribe, referring to "the basin," has forgotten about the "waterfall." The chapter has topical similarity with Chapters 36–40 but is written and illustrated with very inferior techniques. The "Paniscus" mechanism (as well as the "Owl" mechanism, note 582) may be interpreted as a crude, rotary disk valve.

[581] The reference may be to Chapter 58 (sound-effects of song-birds) or to Chapter 59 (sound effects of a dragon drinking from a basin).

without interruption unless you want one. You can make a well-shaped owl, who may sit in front of the birds. Various places can be chosen for him. If the birds shall sing, the owl must not look at them but turn his back to them; the birds then sing until one turns him around. Here is the construction of this apparatus.[582]

Let [invertable] siphons be prepared for this water course, similar to the Egyptian siphons. Let them be made of silver or copper or gilt metal, in

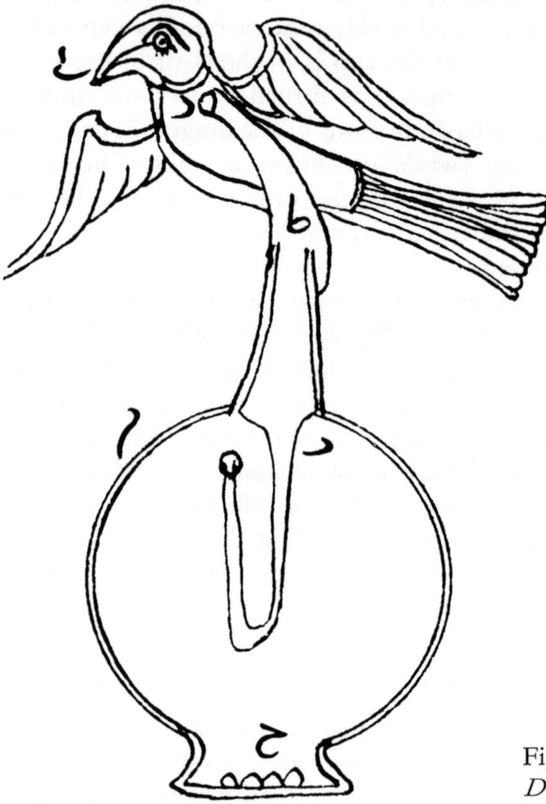

Figure 60. A. S. 3713, fo. 75 a (bottom). *Drawing of Chapter 60.*

order that the water may not corrode them; that impurities may not attach themselves; and that the device may not gradually deteriorate and change its sound. Let their number correspond to the number of the birds. Let some be pierced on their sides, where they dip into the water; change this in others to obtain different sounds. When all is prepared, immerse it in water,

[582] About the owl, also see Chapter 61 a. The exact construction of drawing 60 in **A** is also found in Ibn Musa, drawing 29.

at the proper place. Each bird carries a pipe, whereto the end of the siphon is connected, as shown in the figure. The vessel [specially made for each bird] is α, its mouth β, its bottom γ, the [upturned] siphon δ. The vessel is placed in water and is submerged until the water is at the horizontal line $\varepsilon\zeta$ [of the free end of the siphon]. The siphon head η enters the bottom end of the bird's pipe ϑ. The beak is ι.

When water enters mouth β, air leaves at γ, through η, and enters pipe $\vartheta\iota$. Sound then comes from the beak of the bird, as a whistle is provided at $\varkappa$ in the bird throat. Thus a strong sound is made, since abundant air issues from the vessel as the vessel is submerged. When all air has been driven out of the vessel, one must pour out the water which it contains. One can then submerge it again and repeat the operation.

So much for one such vessel. We can now construct devices with several of them. We use suitable means for submerging them in proper ways and for letting them rise again to automatically refill with air, without any person performing this operation which results.[583] Here is the figure.

[Chapter 61. Whistling Wheel I and Details of an "Owl" Valve][584]

[a]

Construction of a beautiful whistling wheel. Prepare a wheel of wood or copper, having a certain depth, similar to the irrigating tympanum.[585] Let it have two ells diameter. Side plates $\varkappa\lambda$ turn with it. In their outside portions, σ, all around the wheel, these plates are closed. Orifices like those of bladeless tympana are formed on the outside, providing one of the sides of said closed part. On the other side this part has a square-shaped closure, which contains hollow compartments, for receiving the water. Orifices ζ are provided as inlets for these hollow compartments. The surfaces inside υ are closed and provided with [outlet] openings o. Centrally, the wheel has a [shaft] portion ϱ having a diameter of one-third the diameter of the wheel. The [shaft of the] wheel is well supported on strong posts.

---

[583] One such mechanism is shown in Chapter 61a.

[584] The exact construction of this whistling wheel also occurs in **B,** p. 44. It also occurs in the notebook of Fontana[69], fo. 66r.

[585] Vitruvius, X, iv, 1,2. A. P. Usher, *A History of Mechanical Inventions*, Harvard, 1964, pp. 162f. believed that the present construction has significance for the development of water wheels *as power generators*. In any event, as noted by Drachmann, pp. 64f., it shows an *acoustic* development, based on the more elementary Chapter 60.

When all this has been prepared, which must be done well, place the wheel in a vessel filled up to level $\varepsilon\eta$. Above, provide a channel which pours water into hollow openings ζ. The wheel is balanced and of uniform weight on each side. When one side is filled with water, it outweighs the other and the wheel turns. As it turns, empty parts fill. This is done at ζ. As one side of the wheel accordingly dips into the water, air is intercepted. As the water flows and reaches the hollow spaces, the air in these hollow spaces [is driven through pipes, ending at o, wherefrom it escapes and] whistles. As it is carried below the water level, it is expelled with a sound. Meanwhile other parts descend and the process repeats itself. When one [compartment] stops whistling, another begins.

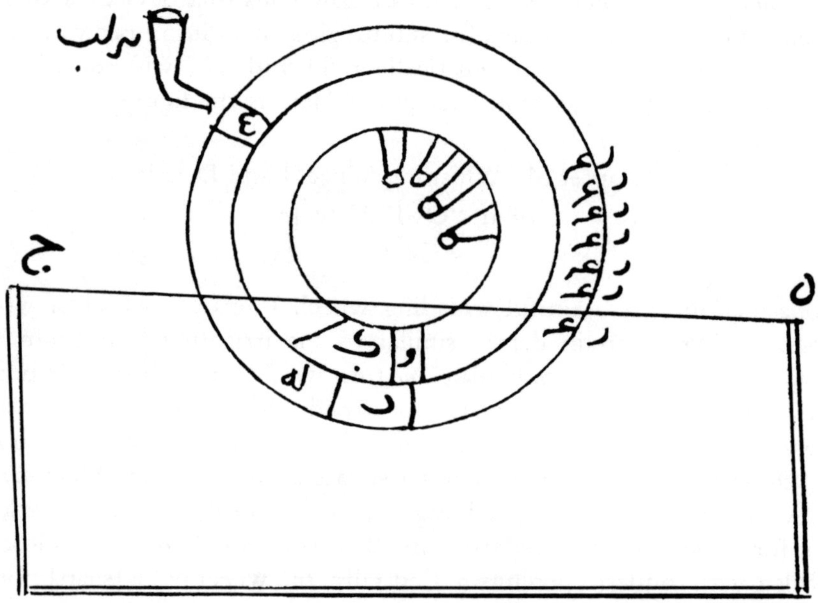

Figure 61. A. S. 3713, fo. 77a (bottom). *Drawing of Chapter 61.*

When the water-receiving compartments rise above the water [level $\varepsilon\eta$], the water leaves them and the peripheral compartments then rise empty. They outweigh those where the sound is produced, since they are larger and their distance from the center is greater.[586]

[586] The author may refer to *negative* "outweighing." Carra de Vaux says, somewath cryptically: "This is the announcement of the law of the lever."

If different sounds shall be produced by different sides, different hollow spaces can be formed. They can be as described, for example four of them, [but] let the hollow devices be different to produce different sounds. They whistle so long as the water flows into the wheel and it turns.

[b]

If they shall not sing any more, turn the owl which we have mentioned. Form a [vertical] tube, divided in its depth into two halves $\gamma\beta$. It can be round or square.[587] Above it let there be a circular plate,[588] which covers [one part of] the round tube, is concentric with it, and turns [with the owl on it] on a diametrical plate [standing vertically] in the middle [of the tube]. Inside,[589] a [spout] member $\varepsilon\zeta$ is soldered to the tube. The water which comes through β is directed to $\varepsilon\eta$ and the channel which leads to the wheel, [but] if you turn the owl the [spout] member no longer receives this water. The part ε is on part γ, and γ is below the center. This the arrangement. Here is the figure.

[Chapter 62. Whistling Wheel II][590]

Construction of another whistle wheel. One can build a water wheel [which whistles by] lifting water from a stagnant pool, not from running water. One builds a water wheel of copper with 1 ell diameter and a marginal width of 1 span. It also has two [inner,] circular margins distant from one another by a similar width. The wheel is α, the [outer] rim of said width is β; the other circles are $\gamma\delta$. On [the rim β of] the wheel, uniform cups are [pivotally] mounted in space $\beta\gamma$ between the two edges. These cups, ε, are curved as shown and have openings ζ. In the next [inner] part there are straight [-walled] chambers η of the indicated form, with openings ϑ turned opposite those of the curved cups. The wheel rests on axles in a square unit and has its lower part dipping into a waterfilled vessel. The water level is $\iota\varkappa$.

When a man turns the wheel in the direction of the open curvature of the rising cups, these cups are full of water as they rise to level $\lambda\mu$. They are [then] emptied [into chambers η, which are provided with central whistle outlets].

---

587 The words "or square" seem to be in error.

588 A *half*-circle seems to be meant.

589 The scribe seems to mean: below the [half-]circle.

590 The reference letters, in the schematic and incomplete drawing, do not agree with the text. There is a related water wheel drawing of Leonardo (C. Pedretti, *Studi Vinciani*, Genoa, 1957, pp. 26f., 34–42); it may be based on suggestions from Gothic variants, now lost.

If one pours water from above into the cups, [which then need not be pivoted; also the man driving the wheel is not then needed; the inner chambers can then receive stagnant water,] these cups outweigh and raise the filled inner chambers, as they are on a larger circle. The device should be built for a certain weight of water, poured into these cups at level $\lambda\mu$, according to a proportion between the water taken up and the water poured into the curved cups. A container on the larger circle outweighs the one on the smaller circle [which holds an equal weight of water]. The wheel [thus] turns as water descends and moves it.

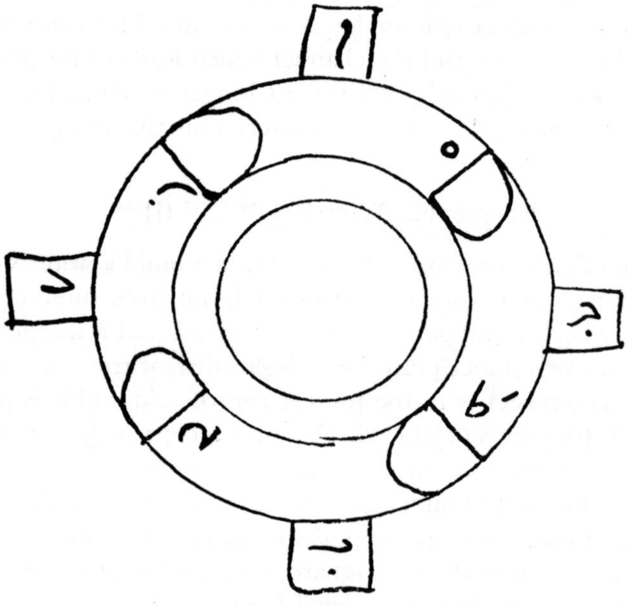

Figure 62. A.S. 3713. fo. 78b (middle).
Drawing of Chapter 62.

This was the description of how to build this device. It is one of the admirable constructions, as its water is not changed or diminished. Here is the figure.

[Chapter 63. Whistling Wheel III, with Concealed Drive]

Another device: a water wheel for ablution and purification, near a *mosque* or temple. This device is similar to the one that we have described, but the

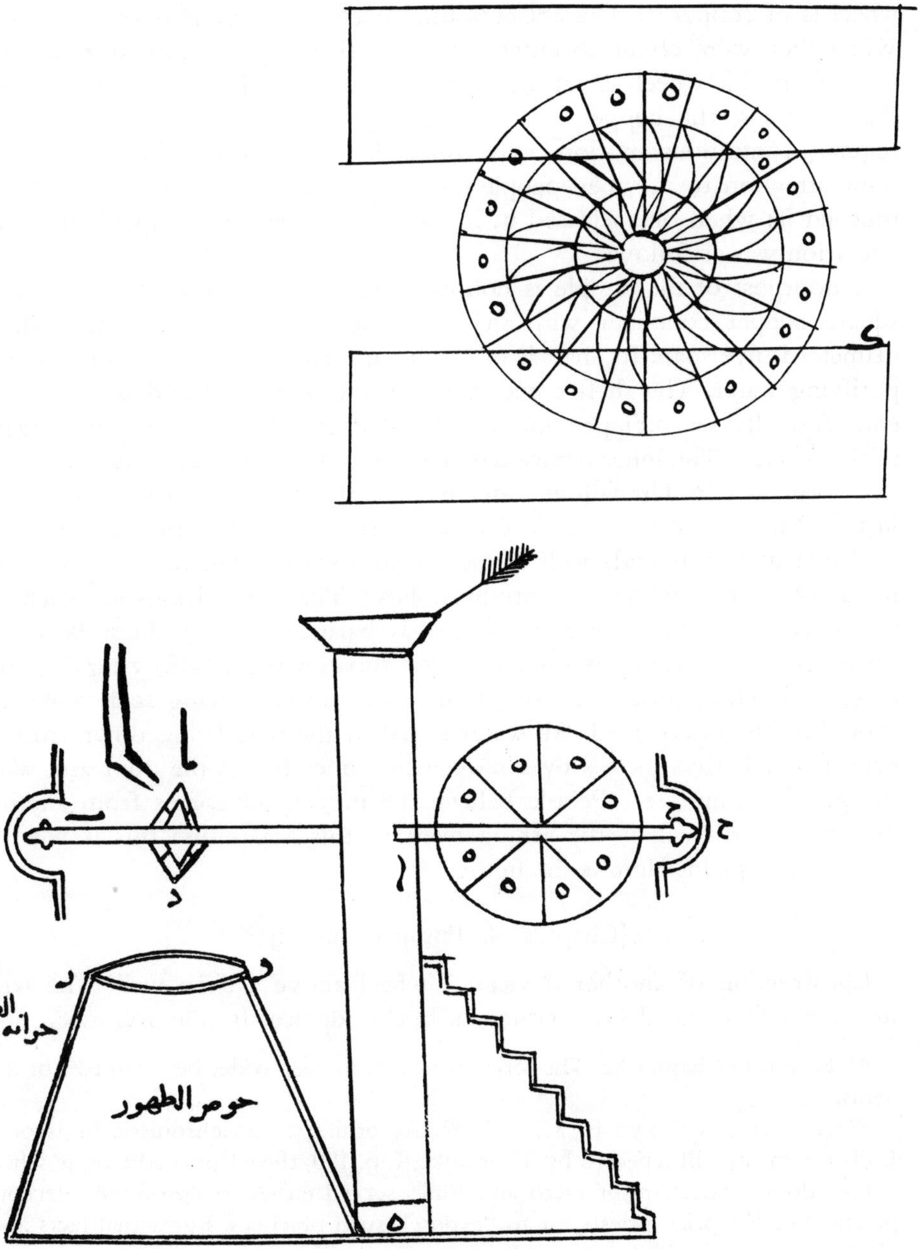

Figure 63. A. S. 3713, fo. 80a (full page) and b (top).
Drawings of Chapter 63.

wheel is of copper.[591] The ancients used many constructions of this type.[592] When they were about to enter a temple, they sprinkled their robes with water from this wheel. Then they put their hands to the wheel; they believed that touching the copper had a purifying effect. The wheel turned with regular, continuous motion, and whistled. The motion and the whistling drew attention to it, when people entered the temple. It stopped when one touched it; when one released it, it started again and turned as before. The operation was as follows.

A doorpost of the temple is hollowed out at a certain place, to receive a square copper container with the wheel in it, turning well; the wheel's diameter is five span. Its shaft is of copper and has one end near the vessel for purifying water. The shaft comes through the door post and is held at two ends. A small reservoir is provided at α; the shaft is β, the outer and purifying part of the wheel γ. The inner device has rim δ. The little cups are mounted around the wheel, at ε.[593] The falling water is concealed in the door-post. It collects in the chambers of the wheel. The water is dispensed from the wheel at ζ.

The shaft has its ends well supported in a square structure specially built for it, of copper, which is centrally hollow. These provisions are such that the wheel turns easily and its weight is balanced on all sides. When the water from the channel is poured in the cups on the wheel, weighing them down, the wheel turns. The supply of water for producing such movement should not be interrupted. All water raised by the wheel runs down from the cups again. It thus moves by a movement concealed in the door and which the people cannot see. People believe the movement comes from anything but the water. Such is the arrangement of this apparatus: this is what we wanted to explain. Here is the figure.

[Chapter 64. Pump Proposal][594]

Construction of another device. A wheel above a deep well. The wheel lifts water from far down, without a bucket device. It is as follows.

---

[591] So it is in Chapter 62. The scribe is unable to keep order between two or three items.

[592] A variant is shown by Hero, I, 32. According to anachronistic ideas of the Enlightenment, still repeated by Thorndike, I, p. 190, the automated temple wheels, temple doors and altars of Hero and Philo served either to "promote" deceptive practices of Egyptian priests, or to "expose" such practices. For actual facts about Egyptian priests in Philo's time: Rostovtzeff, pp. 723, 884, *passim*.

[593] As in Chapter 62.

[594] Drachmann, pp. 6f., rightly questions its operability. It seems to be based on an inept attempt to improve over the Ctesibian pump.

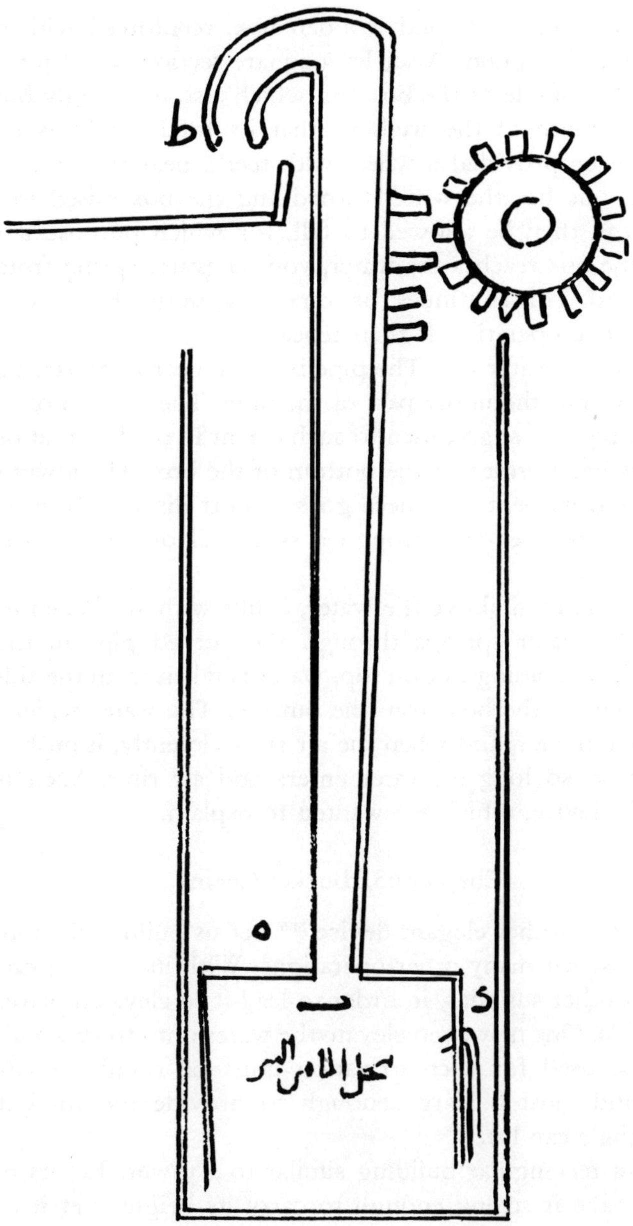

Figure 64. A. S. 3713, fo. 82a (full page). *Drawing of Chapter 64.* Legend in bottom: ب (β) "The Place of the Water in the Well."

Let there be a strong and solid wooden box, reenforced with beams and tightened with tar all around. Also, let a square-section wood pipe be made, which starts in the middle of the box and which also is strongly built; its top rises above the mouth of the well, so that its total height is one fathom higher. Let there be provided a wheel with teeth, near the tube's end.

To lift water, one has the wheel turned and the box raised to above the water level. It can then be allowed to fall, for which purpose it has a lead weight. When the box reaches the water, you see water spring from the pipe, with a strong wind. This continues for some time, until all air has been driven out of the box. The operation is then repeated.

The box above the water is ε. The pipe is ζ, the gear η. Square-section gear teeth ϑ are nailed into the upper part of the pipe. The bottom of the box has an orifice ι. The pipe must go down to such extent in the box that only a small distance remains between it and the bottom of the box. The lower end of the pipe is curved and its bent end piece goes a short distance from the roof of the box. Another pipe enters through a side wall of the box to reach the water.

When the box is raised above the water, it fills with air. When it is thrown in the water, the water springs through the curved pipe in the box. To prevent the air from coming out on top, water is taken from the side, through the pipe which enters the box from the outside. The water which enters the box then is above the air, and when the air rises violently, it pushes the water out. This goes on so long as water enters and air rises. Such is the construction of this device, which we wanted to explain.

[Chapter 65. Bucket Chain]

Construction of another elegant device.[595] Let us build still another device, which one can use for many other operations. With its aid one can lift water from rivers and other supplies, in order to lead it to elevated places, to water gardens and fields. One may even elevate the water onto towers and fortresses. The river to be used for such irrigation must be running swiftly toward lower places, and must be large enough to provide the amount of water which this machine can lift.

Let us build a rectangular building similar to a tower. Let its proportions be such as to make it strong enough to carry its height. Let it be removed

---

[595] Compared with Vitruvius, X, iv, 4, this description is the work of a most competent, ancient or medieval engineer. The Arabic drawing and its reconstruction by Carra de Vaux are technically worthless; the reconstruction by Beck is much better. Also see White[315], pp. 103–109.

some distance from the river, so that the mass of water of the river will not flood the area where the water is raised. Let front and back of the building be spaced to bring the lifting structure to the [proper] place.

Let a wooden floor be placed on these foundations, and the water be led over this floor, which rests on masonry. Let a channel be dug from the river to this building, $1\frac{1}{2}$ fathoms deep, below the horizontal, that is, below the water level in the channel. Let the walls of this channel be built solidly, with a bottom made with lime and cement, carefully worked, and let it extend to the basin. Let the walls of the basin have 6 ells length and such distance in width as to closely fit the irrigation device.

This device comprises a strong cross-shaft with pullies of two ell diameter thereon; each end of the shaft being housed and held in a square bearing member, wherein it turns easily. The entire device is very solidly fitted because its motions are strong. At the height of the building another strong shaft is provided, which is similar to the lower one.

The wheel for irrigating motion is in the middle of the shaft. Its diameter is four ells. Let a triangular structure be made of copper, with sides of such length as results when they correspond to the radius of the water wheel.[596] The structure has one ell width. On the two end pieces of the shaft are two pulleys of uniform size similar to the aforesaid lower pulleys. The central water wheel, with triangular structure and pulleys, is fixed to the shaft. The lower water wheel is $\alpha\beta$, on shaft γ with pulleys $\delta\varepsilon$; the upper wheel is ζ, with triangular structure η, on shaft ϑ with pulleys $\iota\varkappa$.

Also needed is an iron link structure, to be driven by the pulleys.[597] Let it resemble a vertebral column and let its length be such that, hanging from the triangular structure, it comes within one ell of the basin's bottom. Let the segments also be one ell from one another. Let it be articulated by iron pins. The device is o, the pins λ. Let rectangular cups be made of copper or wood, pinned to this device on the inner side [of the chain that they form]. Let the device be placed about the triangular structure, with the cups pinned to the device.

When the shafts are turned with force, the triangular structure turns and the cups, full of water, rise. Each part of the link device, carrying a cup, necessarily falls on the angles of the triangle, so that, when the wheel tilts the

---

596 Probably meaning sides normal to three equidistant radii and intersecting these radii near their ends, so that the sides are 6 ells long.

597 That is, a water-bucket chain driven by two power chains. The only reference to iron, and to a chain, in the entire book.

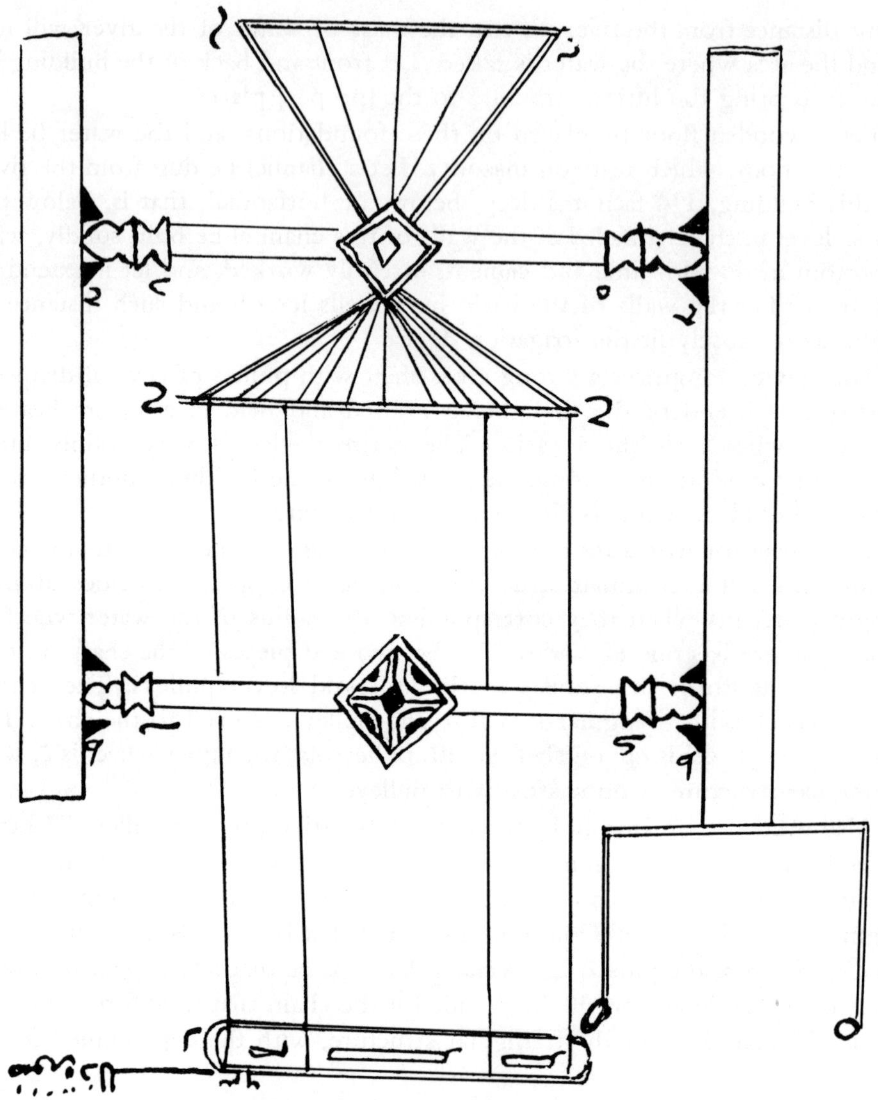

Figure 65. A. S. 3713, fo. 84b (full page).
Drawing of Chapter 65.

full cups, they empty above point *ν*.[598] Here one places a channel, which receives this water and carries it off to the duct on masonry pillars, as we have mentioned.[599]

We must still explain how the shafts are moved automatically to raise the water from the aforesaid channel. Let spokes be made, which end in water wheel blades. Let them be solid and strong and let them be so arranged that, when the blades receive a full impact, the lower shaft turns with very strong motion. When the lower shaft moves with force, and even vibrates, the upper one also moves by virtue of the chain unit whereon the cups are mounted.

Four cups are fitted for each segment of the apparatus; each has two *kouz* volume.[600] The speed depends on the abundance or rarity of the water. It should be noted that of the two water wheels, that with the triangular structure should be the largest. If the device is strong enough to lift five cups, it can be 60 ells high. The rising of the cups will be easy if the basin is deep enough to submerge them and fill them. Excess water can be deflected to another channel, which may go to a lower area. This is how this device is built. Here is the figure.

---

598 Contrary to Carra de Vaux, this clearly specifies that all cups empty above this point.

599 In manuscripts **A** and **B**, the duct and pillars have *not* been mentioned, up to here.

600 It appears that 1 *kouz* equalled about 2½ cubic ells. This measure is not noted by Hinz[103].

The „*Extracts of Irun*"

For the last hundred years it has been fascinating to historians of technology to read about the manuscript from the Arabic middle ages, which suggested that Philo had described ancient constructions, including water-driven and water-lifting wheels and levers, of types still used in the orient. This Arabic document (Eastern Manuscript 954 of the Bodleian Library in Oxford, entitled "*Extracts of Irun, from Filun and Archimedes the Greeks,*") is here designated as version **B.** It remained fascinating when it was found partly confirmed by the Arabic copy of Philo's *Pneumatica,* and by the reported remark of AlKindus of Egypt, describing Philo as a man "learned in water wheels and mills" as well as other "ingenious devices."

However, the ancient evidence is silent about Philonian constructions or descriptions of "water wheels and mills." It rather shows Philonian descriptions and illustrations of waterlifting siphons and the like. The Arabic medieval reports are therefore in need of critical review: a historic doctrine requires documentation rather than fascination. Until the time when the authentic text of AlKindus may be rediscovered, this critical review must concentrate on the "*Extracts of Irun.*"

The booklet containing these *Extracts* was acquired, through unknown channels, by Bishop Narcissus Marsh, in Ireland. It was his document No. 669, and came as such to the Bodleian Library. It contains only the *Extracts.* The booklet is a slender one, of quarto size, on paper, executed in black and red ink, which still is very brilliant. The date of this copy is unrecorded; in view of internal evidence it can hardly be much earlier than 1400, if indeed it is that old. The underlying tradition no doubt is much older.[601]

Chapters from **B** were published by Carra de Vaux in his edition of **A.**[602] He recognized **B** as a late and confused copy of earlier material, also reflected by **A.** The text edition of Carra de Vaux was superb, in this case as generally; however, his drawings and some of his conclusions were questionable. As rightly noted by Drachmann, the drawings of **A** and **B,** as given in the French edition of 1903, are graphically invalidated by the fact that they

---

601 Carra de Vaux, p. 294.

602 After Carra de Vaux, Advance Notice, pp. 28–38.

are redrawn in a modern style and without making it possible for the reader to consider their own, authentic drawing techniques, and their relationships with other and more clearly datable drawings. Beck, a most qualified observer, also questioned large parts of the Carra de Vaux attribution of manuscript **B** to Philo. The contents of this manuscript were then republished, with more adequate copies of their authentic illustrations, by Wiedemann-Hauser. In 1921, Carra de Vaux himself reconsidered his former attributions and concluded that the Arabic versions and copies are not purely Philonian, but that they "contain, or have acquired, other important fragments of ancient mechanics."[603] This was the last conclusion of Carra de Vaux, and may be accepted as a most plausible opinion.

Evidently, we should only accept copy **A** as Philonian to such extent as it contains pneumatics authenticated by other evidence. To such extent, the parallel contents of **B** may also be truly Philonian. Other parts of **B** provide a rather confused copy, reflecting Philonian and non-Philonian contents of ancient mechanics.

For purposes of this edition I compared manuscript **B,** in Oxford, with a complete set of photo-reproductions of the pneumatics chapters in **A,** and with the principal copies of the Latin set **C** to **Z.** On such comparison, the mill works and pump work in **B** appeared quite foreign to the Philonian tradition, in graphic technique as well as contents. Equally foreign to all real pneumatics of Philo are the few corresponding parts of **A** (Chapters 64 and 65).

As shown in Chart 3, above, Codex **B** uses a dedication to "Ristun." No doubt this is traceable directly to "Ariston" and thereby to Philo, but it leaves the question open to what extent the attribution to Philo is valid.

Since the opinion of Carra de Vaux, 1921, has remained rather unknown and his opinion of 1903 continues to be cited with all too little criticism, some critical notes about it are necessary. In his edition of 1903, Carra de Vaux believed to recognize a number of collections and fragments, in the *Extracts*; however, he added further confusion to this confused manuscript. His analysis was as follows:

Folios 2 to 6, under the title "First *Collection of Apparatus and Machines*:" two pumps and a four-liquid vessel.

Folios 6 to 18, under the title "*Book about Wheels which move by themselves*:" a collection of perpetual motion devices.

Folios 19 to 21: a fragment of water clocks.

---

[603] Drachmann, pp. 41f., 45; Beck, p. 77; Wiedemann-Hauser on Irun's Extracts, pp. 147–154; Carra de Vaux, *Penseurs*, II, p. 171.

Folios 22 to 25: another fragment on water clocks.

Folios 26 to 48: the anonymous *Collection.*

Folios 49 to 92: the "*Book of Pneumatics of Philo.*"

The translation of titles which are reported in the *Extracts* is quite correct. The analysis calls for the following remarks.

The manuscript has only 48 folios. Wherever the word "folios" occurs in the Carra de Vaux analysis, it must be replaced by "*pages*". It must further be noted that the pages must be counted; the book is unpaginated. In the present edition the verso of the title page, at the right end of the book, is counted as folio 1a or page 2; it has the title "First *Collection* . . .," noted by Carra de Vaux.

The Wheels which Move Themselves, on pages 6 to 18, are *automata,* not "perpetual motion devices." The contents of pages 19 to 25 are not "water clocks", but are *swapes* (water scoops on levers). The subtitles found on pages 26 and 49 are misleading; the excerpts from manuscript **A,** cited as the "Book of Pneumatics by Philo," begin on *page 39.*

The actual contents of the manuscript are quite simple:

A first part comprises pages 1 to 38: a series of non-pneumatic machines such as water pumps and bucket systems of the swape or winch or mill types. The winch type water lifters are related with machines known from Vitruvius.

The second and last part consists of pages 39 to 96. It is an excerpt from manuscript **A,** or rather from a related, unknown, Arabic prototype, wherefrom it also takes a few variants. Up to page 50, this excerpt illustrates and describes machines of **A,** Chapters 58 to 65; then it continues with machines of **A,** Chapters 28 to 41.

The drawings of this second part, pages 39 to 96, are almost identical with the corresponding ones in **A,** published here. Those of the first part, pages 1 to 38, illustrate machines entirely foreign to those of **A,** and illustrate them in a schematic style and drawing technique, different from those which are used in the first two scores of chapters in **A.** Their content and style are equally foreign to the Latin Philo, manuscripts **C** to **Z.**

The scribe of **B** does not affirmatively attribute this first part to Philo.[604] The internal evidence, provided by the suggestions in titles and subtitles

[604] He attributes certain developments of swapes and water bucket mechanisms, quite plausibly, to "the Ispahanians" or Persians; Carra de Vaux, p. 223. Ispahan, the ancient Aspadana, was the medieval capital of Persia. In Philo's time it was the center of local, horse-raising activities.

contributed by this scribe, is not a good reason for so attributing this part. The scribe *may* have believed, and his predecessor AlKindus reportedly *did* believe, that Philo was "learned in water wheels and mills." The scribe suggested this opinion, which his Arabic predecessor had stated more explicitly, to his French editor, in 1903, but the suggestion became doubtful to this editor and to others, during the years which followed. It was a suggestion devoid of foundation in the ancient record. At present we may conclude that Manuscript **B,** the *Extracts of Irun*, adds nothing of validity to the Greek and Gothic fragments, the Latin-Italian record, and the Arabic edition of Philo's *Pneumatica*, concerning the contents of the *Pneumatica*.

Conclusions (A summary of the evidence, presented in this edition)

Observers and thinkers of archaic and classic times provided Philo with a body of factual, pneumatic knowledge, and with a mass of pertinent dialectics. In his *Pneumatica*, Philo reflected the earlier observations; he disregarded the dialectics. He presented the facts as part of a *Mechanical System*, and tried to explain them by a pneumatic theory. Most importantly, he developed a way of teaching the facts by visual observation.

Most clear, most original, and most famous among Philo's followers is Hero, a few centuries after him. The two men shared a fundamental, physical view, in agreement with the view of Democritus and Strato: they saw air and the other substances as composed of particles, with *vacuum* disseminated between them. Their view rightly, and in Hero's case rather explicitly, rejected the Platonic and Aristotelian postulates of a *continuum*.

The specific view of Philo, as perhaps the one of Strato but not, it seems, the one of Hero, explained the pneumatic lifting of liquids by a cohesion of the atoms, mingled with the voids. Philo tried to prove this lifting, and this explanation, by some of his demonstrations. He developed these from devices which had been designed by his teacher Ctesibius and by earlier predecessors. Some such devices may still be found as archaeological artifacts.

According to present-day physics, the theory of cohesion was in error. However, the ancient demonstrations, shown in connection with it, may be seen as one of the closest approaches, made in any part of ancient science, to the modern experimental method; they used and forcibly suggested full reliance on observational induction and proof. Predecessors of Philo, and his own *Belopoiica*, already had made significant use of such induction; his *Pneumatica* may be seen as the first attempt to use it systematically.

The pneumatic demonstrations were popular. They mixed entertainment with enlightenment. Artisans as well as scribes were kept familiar with the pneumatic constructions, as a need for demonstrating devices persisted. The continued building of the devices may be expected, in turn, to lead to new ideas and new techniques; and indeed we find such developments, mainly in the Persian-Arabic successors' notebooks. In a modest way, such developments may also have continued in the west, even during the dark ages of physical science. However, theoretical pneumatics, of either classic or

Hellenistic type, was substantially forgotten, until the western and mainly the Sicilian renewal of Greek science.

Even after this renewal, the progress of pneumatic theory was slow. Earlier manuscripts of Philo's work had become so old and so worn as to be largely unintelligible. Pages and parts of them were misplaced or lost, and even the most dedicated copyists could hardly make sense of the work, in a number of chapters. Earlier copyists had frequently confused variants and extraneous additions with authentic copies; had misunderstood the old dedications, manuscript titles, chapter titles, and sequences of chapters; and had made it hard indeed for their followers to gain real access to Philo's work. In response to these difficulties, eastern editors produced original variants of Philo's work, while western editors produced radically abbreviated versions. There are Gothic fragments, some of them retaining Philo's name or being appended to his book; they indicate that one of the western traditions had lost all but three "hermetic" chapters, once attributed to Philo.

A result of these medieval developments may be seen in the sixty-five chapters of the Arabic-Latin tradition of Philo's *Pneumatica*, which are translated in this edition. These chapters appear to be a compilation of two works, both reflecting recollections from Philo. Both are subject to distortions, which must have limited the understanding of the work by medieval scientists and artisans. In particular, the two works are subject to two different types of distortion: excessive abbreviation of the first and excessive amplification of the second.

The first score of chapters, relatively best preserved in the Latin tradition, reflects an important part, but not the entirety, of Philo's original work on pneumatic laws and demonstrations. The text is derived from an abbreviation made in Hellenistic times. The abbreviators rendered the pneumatic theory of Philo and its demonstrations by an illustrated text which can be understood and accepted as a correctly abbreviated copy, in general. At a few points, however, this abbreviation corrupted the original statements; so for example in Chapter 6. The abbreviators also shifted one part, Chapter 4b, to an ill-fitting, early position, where it disturbs the logical structure of the work. In addition, some important, authentically required parts of the work are lacking: according to Philo's own remarks, the book was supposed to deal not only with established laws of pneumatics but also with problems which were new and debatable in his time: the forcible compression of air, and the natural mechanism of "cohesion" between the atoms. These two problems are no longer discussed or explained in the *Pneumatica*.

Philo may have forgotten them, when he wrote the work on pneumatics, or they may have been lost by later editors and abbreviators. Nevertheless, the Latin-Arabic record of the first score of chapters appears, in general, to be a fairly correct account of the pneumatic views, which Philo had developed from earlier observations. In many respects these chapters are confirmed by archaeological evidence.

The other group of chapters, about twice as large as the first, is less clear and less unified than the first. It includes pneumatic propositions more or less similar to the preceding ones, but it also includes propositions utterly irrelevant to pneumatics, as well as chapters in very inferior form. The chapters of this group may stem from Philo or from Hellenistic or Roman-Byzantine followers, but hardly from his book on pneumatics. Some of the followers seem to have belonged to the Celtic, Carolingian and Gothic cultures in the areas of Western Rome, and to the Byzantine, Persian and Arabic cultures in the east. Some of the proposals relate to bath heaters, censers and candle units, and seem to have developed into a small, "hermetic" compilation. Two final chapters in the principal, Arabic manuscript, as well as a number of chapters in Arabic excerpts and variants, deal with water wheels and water supply systems for aqueducts. According to ancient evidence they are not attributable to Philo's *Pneumatika*.

An archetype of the sixty-five chapters probably was in existence some centuries after Hero's time. It was paralleled and followed by excerpts, both east and west. Then it was followed by Latin and Arabic prototypes, in the XII. and XIII. centuries. The eastern tradition influenced the western one, and *vice versa*.

Because of an eastern influence, shown by the beginning line of the Latin Philo text, it has been assumed that this entire text is only a translation from an Arabic prototype. This opinion has been held for the last hundred years; it also has support in a tradition of three hundred years ago. However, according to many bits of evidence, the Philo traditions have been numerous and complex, and the old opinion about the Latin text was over-simplified.

In any event, many copies of the Latin Philo were produced and studied, and in some cases they gave rise to new, original proposals. In the Italian Renaissance, their contents became available, again, to practicing and experimenting masters of handicraft, as well as scholars.

By then the *Pneumatica* of Hero had been rediscovered. It contains explicit demonstrations of extended vacua, and therefore became more challenging than the *Pneumatica* of Philo, to those who doubted the Aristotelian *horror vacui*. Philo then was eclipsed by his great pupil.

However, he was not entirely forgotten. One of his chapters still gave rise to the renewal of the old thermoscope in form of the modern thermometer, a most important instrument of scientific observation. Galileo was one of those who contributed to the modern development of this instrument, and to other renewals of Philonian views. Not all the views were correct, but their spirit supported the great man in his struggle for observable, physical truth, truly demonstrated. Only a few more years were then needed, in Galileo's circle, for some of the great achievements of modern science: the rediscovery of atmospheric pressure, the re-interpretation of atmospheric height, and the discovery of the Torricellian Vacuum.

Bibliography

of Philo's Works and their Editions,
of Major Background Works, and
of Major Works Showing his Influence and Reputation

AlKindus (of Egypt), *Automata and Wondrous Machines*. (Written in the IX. century. According to Herbelot[69], p. 363, the author died A.D. 868. The work is lost. Latin excerpts: Giovanni Fontana[69], cod. icon. 242, Bayerische Staatsbibliothek, Munich).

—, *Description of Egypt*. (Translated excerpts: E. Wiedemann, Beitrag III in Vol. XXXVII, 1905, pp. 221–233, cited as "AlKindus, ed. Wiedemann." See below, *s.v.* Wiedemann, for full citation of his "Beiträge".)

Aristotle, *The Works*, translated into English, ed. D. Ross, Oxford, 1908–1931. Cited as "Aristotle."

Beck, Theodor, "Philon von Byzanz," in *Beiträge zur Geschichte der Technik und Industrie*, II, 1911, pp. 64–77. Cited as "Beck".

—, "Heron des Älteren Automatentheater," in *Beiträge zur Geschichte der Technik und Industrie*, I, 1909, pp. 182–199. Cited as "Beck, Automaten."

—, *Beiträge zur Geschichte des Maschinenbaues*, Berlin, 1899. Cited as "Beck, *Beiträge*."

Carra de Vaux, Baron Bernard, "Le livre de Philon sur les appareils pneumatiques et les machines à eau," in *Notices et Extraits de l'Académie des Inscriptions et Belles-Lettres*, XXXVIII, 1903, pp. 27–40 (Introduction), 41–121 (Arabic transcript), and 122–235 (French translation with modern illustrations). Cited as "Carra de Vaux."

—, "Notice sur un manuscrit arabe . . .," in *Bibliotheca Mathematica*, 3. series, I, 1900, pp. 28–38, cited as "Carra de Vaux, Advance Notice."

—, "Les mécaniques de Héron," in *Journal asiatique*, 9. series, I, 1893, pp. 386–472; II, 1893, pp. 152–269 and 420–514, cited as "Carra de Vaux, Heron."

—, *Les penseurs de l'Islam*, Paris, 1921–1926, cited as "Carra de Vaux, *Penseurs*."

—, and Paul Tannery, "L'invention de l'hydraulis," in *Revue des études grecques*, 1908, pp. 326–340, cited as "Carra de Vaux, Hydraulis."

Daremberg-Saglio, *Dictionnaire des antiquités grecques et romaines*, Paris, 1873–1919. Cited as "Daremberg-Saglio."

Deonna, Waldemar, "Vases à surprise et vases à puiser le vin," in *Bulletin de l'Institut National Genevois*, XXXVIII, 1909, pp. 207–233.

Diels, Hermann, *Die Fragmente der Vorsokratiker*, 10. ed. (by W. Kranz), Berlin, 1960–1961. Cited as "Diels, *Fragmente*."

—, "Über das physikalische System des Straton," in *Sitzungsberichte der Kgl. Preußischen Akademie der Wissenschaften*, 1893, pp. 101–127. Cited as "Diels, Strato."

—, *Antike Technik*, Berlin, 1920.

—, also see his editions of Philo's *Belopoiica* etc.

Drachmann, A.G., *Ktesibios, Philon and Heron*, Copenhagen, 1948 (*Acta historica scientiarum*, IV). Cited as "Drachmann."

—, *The Mechanical Technology of Greek and Roman Antiquity*, Copenhagen, 1963. Cited as "Drachmann, *Technology*."

—, "On the alleged second Ktesibios," in *Centaurus*, II, 1951, pp. 1–10. Cited as "Drachmann, Second Ktesibios."

Haase, Friedrich, "Philon der Byzantier," in Ersch-Gruber, *Allgemeine Encyklopädie der Wissenschaften und Künste*, Leipzig, 1818–1850, III. Sektion, XXIII, pp. 428–432. Cited as "Haase."

Hammer-Jensen, Ingeborg, "Die Druckwerke Herons von Alexandria," in *Neue Jahrbücher für das klassische Altertum, Abhandlungen*, XXV, 1910, pp. 413–427, 480–503. Cited as "Hammer-Jensen."

—, "Ptolemaios und Heron," in *Hermes*, XLVIII, 1913, pp. 224–235.

—, "Die Heronische Frage," in *Hermes*, LXIII, 1928, pp. 34–47. Cited as "Hammer-Jensen, Heronische Frage."

Hauser, F., see Wiedemann-Hauser and Ibn Musa.

Hero of Alexandria, principal early editions. G. Valla, partial ed. of "*De spiritalibus*" in *De expetendis et fugiendis rebus*, Venice, 1501, Book XV, i, nine pages, unpaginated. F. Commandino, ed. of *Spiritalium liber*, Urbino, 1575. B. Baldi, *De gli automati*, Urbino, 1589. *Veteres mathematici*, Paris 1693, see Philo. Bennet Woodcroft, *Pneumatica*, London 1851, re-edited by M. Boas-Hall, New York and London, 1971.

—, *Opera quae supersunt*, I, ed. Wilhelm Schmidt, Leipzig, 1899: Hero's *Pneumatica* (pp. 1–333) and *Automatopoiica* (pp. 334–457, both in Greek and German, both cited here as "Hero;" Rose's text of the Latin Philo (pp. 458–489, in Latin and German); and a separate volume, "*Supplementum*" (pp. 1–181), on the history of the manuscripts and printed editions. An "Introduction" (pp. I–L of the basic volume) adds remarks about the manuscript illustrations.

Ibn Musa, Ahmed, *Das Werk über die sinnreichen Anordnungen*, c. 850, ed. F. Hauser, in: *Abhandlungen zur Geschichte der Naturwissenschaft und Medizin*, I, 1922. Introduction, pp. 1–33, cited as "Hauser on ibn Musa." Illustrated text, pp. 34–173, cited as "Ibn Musa."

—, also see Wiedemann-Hauser.

Krafft, Fritz, "Heron von Alexandria," in: *Die Großen der Weltgeschichte*, Zürich, 1972, II. Cited as "Krafft."

Pauly-Wissowa, *Real-Encyclopedie der classischen Altertumswissenschaft*, Stuttgart, 1894—, cited as Pauly-Wissowa.

Philo of Byzantium, *Mēchanikē Syntaxis*:

—, *Belopoiica*, *Parasceuastica* and *Polioretica*, ed. M. Thévenot and P. De la Hire, in: *Veterum mathematicorum : Athenaei, . . . Philonis et aliorum opera* graece et latine, Paris, 1693, pp. 49–104. The newer editions, for example ed. Diels-Schramm, in *Abhandlungen der Preußischen Akademie der Wissenschaften, Philosophisch-historische Klasse*, 1918, 1919, repeat the pagination of *Veteres mathematici*, which is generally cited here as "Philo, *Belopoiica*."

—, *Pneumatica*, edition of 1870, see Rose;
edition of 1881, see Rochas, de;
edition of 1899, see Schmidt, ed. of Hero;
edition of 1903, see Carra de Vaux;
discussion, 1910, see Hammer-Jensen;
discussion, 1911, see Beck;
discussion, 1948, see Drachmann.

—, also see Index, s.v. *Automatopoiica*, *Eisagoge*, *Limenopoiica*, *Mochlica*, *De quinque facultatibus staticis*, and *De mirabilibus arbitriis*.

Porta, Giovambattista della, *I tre libre de'spiritali, cioè d'inalzar acque per forza dell'aria*, Naples, 1606. Cited as "Della Porta."

Rochas, Albert de, "Traîté des *Pneumatiques* de Philon," in *Revue d'Archéologie*, 1881, cited as De Rochas, 1881; also in his book, *La science des philosophes et l'art des thaumaturges*, Paris, [1912], pp. 195–208 and Pl. XIII, XIV, cited as "de Rochas, 1912."

Rose, Valentin, *Anecdota graeca et graecolatina*, Berlin, 1864–1870, II, pp. 281–314 and Pl. I, II (republished, Amsterdam, 1968). Cited as "Rose."

—, *Aristoteles pseudepigraphus*, Leipzig, 1863. Cited as "Rose, Aristoteles."

Rostovtzeff, M., *The Social and Economic History of the Hellenistic World*, Oxford, 1953. Cited as "Rostovtzeff."

Sarton, George, *A History of Science*, Harvard, 1952–1959. Cited as "Sarton."

—, *Introduction to the History of Science*, Baltimore, 1927–1948. Cited as "Sarton, *Introduction*."

—, *Ancient Science and Modern Civilization*, New York, 1954. Cited as "Sarton on Euclid and Ptolemy."

Schmidt, Wilhelm, "Physikalisches und Technisches bei Philon von Byzanz," in *Bibliotheca mathematica*, 3. series, II, 1901, pp. 377–383, cited as "Schmidt, Physikalisches."

—, "Zur Geschichte des Thermoskops," in *Abhandlungen zur Geschichte der Mathematik*, VIII, 1898, pp. 163–173, cited as "Schmidt, Thermoskop."

—, "Aus der antiken Mechanik," in *Neue Jahrbücher für das klassische Altertum*, 1904, pp. 329–351, cited as "Schmidt on Carra de Vaux."

—, "Leonardo da Vinci und Heron," in *Bibliotheca mathematica*, 3. series, III, 1902, pp. 180–187, cited as "Schmidt on Leonardo."

—, "Heron von Alexandria im XVII. Jahrhundert," in *Abhandlungen zur Geschichte der Mathematik*, VIII, 1898, pp. 206ff., cited as "Schmidt, XVII. Jahrhundert."

—, ed., see Hero, *Opera*.

Singer, Charles, et al, *A History of Technology*, Oxford, 1954–1958. Cited as "Singer."

Thorndike, Lynn, *A History of Magic and Experimental Science*, New York, 1923–1956. Cited as "Thorndike."

Veteres mathematici. See Hero of Alexandria, principal early editions, and Philo of Byzantium, *Belopoiïca*.

Vitruvius, *De architectura*. Cited in English after the translation by M.H. Morgan, Harvard, 1914.

Wiedemann, E., "Beiträge zur [arabischen] Geschichte der Naturwissenschaften", in *Sitzungsberichte der physikalisch-medizinischen Societät zu Erlangen* (Reprinted, Hildesheim, 1970). Cited in the following form: "Wiedemann, Beitrag III in Vol. XXXVII, 1905."

— und F. Hauser, "Über Trinkgefäße und Tafelaufsätze," in *Der Islam*, VIII, 1918, pp. 55–93 (cited as "Wiedemann-Hauser on AlJazari, Drinking Vessels") and pp. 268–291 (cited as "Wiedemann-Hauser on Ibn Musa, Drinking Vessels").

— und F. Hauser, "Über Vorrichtungen zum Heben von Wasser in der islamischen Welt," in *Beiträge zur Geschichte der Technik und Industrie*, VIII, 1918, pp. 121–147 (cited as "Wiedemann-Hauser on AlJazari, Water Lifters") and pp. 147–154 (cited as "Wiedemann-Hauser on Irun's *Extracts*").

—, also see Ibn Musa, ed. F. Hauser.

Index

of Philo's Background; his Life, Work and Concepts; and the major Copies, Editions and Variants of his Work.[605]

---

605 NOTE: names and works of additional successors and editors, as well as commentators, are cited in the Bibliography and in the footnotes.